This report contains the collective view
ternational group of experts and does no
represent the decisions or the stated po
United Nations Environment Programme, the International Labour Organisation, or the World Health Organization.

Environmental Health Criteria 124

LINDANE

Published under the joint sponsorship of
the United Nations Environment Programme,
the International Labour Organisation,
and the World Health Organization

First draft prepared by Dr M. Herbst, International Centre for the Study of Lindane, Lyon, France and Dr G.J. Van Esch, Bilthoven, The Netherlands

World Health Organization
Geneva, 1991

The **International Programme on Chemical Safety (IPCS)** is a joint venture of the United Nations Environment Programme, the International Labour Organisation, and the World Health Organization. The main objective of the IPCS is to carry out and disseminate evaluations of the effects of chemicals on human health and the quality of the environment. Supporting activities include the development of epidemiological, experimental laboratory, and risk-assessment methods that could produce internationally comparable results, and the development of manpower in the field of toxicology. Other activities carried out by the IPCS include the development of know-how for coping with chemical accidents, coordination of laboratory testing and epidemiological studies, and promotion of research on the mechanisms of the biological action of chemicals.

WHO Library Cataloguing in Publication Data

Lindane.

(Environmental health criteria ; 124)

1. Benzene hexachloride – adverse effects 2. Benzene hexachloride – toxicity
3. Environmental exposure 4. Environmental pollutants
I.Series

ISBN 92 4 157124 1   (NLM Classification: WA 240)
ISSN 0250–863X

Publications of the World Health Organization enjoy copyright protection in accordance with the provisions of Protocol 2 of the Universal Copyright Convention. For rights of reproduction or translation of WHO publications, in part or *in toto*, application should be made to the Office of Publications, World Health Organization, Geneva, Switzerland. The World Health Organization welcomes such applications.

The designations employed and the presentation of the material in this publication do not imply the expression of any opinion whatsoever on the part of the Secretariat of the World Health Organization concerning the legal status of any country, territory, city, or area or of its authorities, or concerning the delimitation of its frontiers or boundaries.

The mention of specific companies or of certain manufacturers' products does not imply that they are endorsed or recommended by the World Health Organization in preference to others of a similar nature that are not mentioned. Errors and omissions excepted, the names of proprietary products are distinguished by initial capital letters.

Printed in Finland
91/8956 — Vammala — 5500

CONTENTS

1. SUMMARY AND EVALUATION; CONCLUSIONS; RECOMMENDATIONS 13

   1.1 Summary and evaluation 13
       1.1.1 General properties 13
       1.1.2 Environmental transport, distribution and transformation 13
       1.1.3 Environmental levels and human exposure 14
       1.1.4 Kinetics and metabolism 15
       1.1.5 Effects on organisms in the environment 16
       1.1.6 Effects on experimental animals and *in vitro* 17
       1.1.7 Effects on humans 19
   1.2 Conclusions 19
       1.2.1 General population 19
       1.2.2 Subpopulations at special risk 20
       1.2.3 Occupational exposure 20
       1.2.4 Environmental effects 20
   1.3 Recommendations 20

2. IDENTITY, PHYSICAL AND CHEMICAL PROPERTIES, ANALYTICAL METHODS 21

   2.1 Identity 21
       2.1.1 Primary constituent 21
       2.1.2 Technical product 22
   2.2 Physical and chemical properties 23
   2.3 Conversion factors 23
   2.4 Analytical methods 24
       2.4.1 Sampling 24
       2.4.2 Analytical methods.............................. 24

3. SOURCES OF HUMAN AND ENVIRONMENTAL EXPOSURE 26

   3.1 Natural occurrence 26
   3.2 Man-made sources 26
       3.2.1 Production levels and processes 26
           3.2.1.1 Manufacturing process 26
           3.2.1.2 World-wide production figures 26
       3.2.2 Emissions 27

		3.2.3	Uses	27
		3.2.4	Extent of use	27
		3.2.5	Formulations	28
4.	ENVIRONMENTAL TRANSPORT, DISTRIBUTION, AND TRANSFORMATION			29
	4.1	Transport and distribution between media		29
		4.1.1	Volatilization	29
		4.1.2	Precipitation	29
		4.1.3	Movement in soils	30
		4.1.4	Uptake and translocation in plants	32
	4.2	Biotransformation		33
		4.2.1	Degradation	33
			4.2.1.1 Degradation under humid conditions	33
			4.2.1.2 Degradation under submerged conditions	34
		4.2.2	Degradation under field conditions	34
		4.2.3	Hydrolytic degradation	37
		4.2.4	Photolytic degradation (laboratory studies)	37
		4.2.5	Biodegradation in water	37
		4.2.6	Microbial degradation (field studies)	38
		4.2.7	Bioaccumulation/biomagnification	40
			4.2.7.1 n-Octanol/water partition coefficient	40
			4.2.7.2 Aquatic environment	40
			4.2.7.3 Terrestrial environment	42
			4.2.7.4 Bioconcentration in humans	43
			4.2.7.5 Field studies	43
5.	ENVIRONMENTAL LEVELS AND HUMAN EXPOSURE			45
	5.1	Environmental levels		45
		5.1.1	Air	45
		5.1.2	Water	46
			5.1.2.1 Rain and snow	46
			5.1.2.2 Fresh water	46
			5.1.2.3 Sea water	47
		5.1.3	Soil	48
			5.1.3.1 Sediment	48
			5.1.3.2 Dumping grounds and sewage sludge	49
		5.1.4	Drinking-water, food and feed	49
			5.1.4.1 Drinking-water	49
			5.1.4.2 Cereals, fruits, pulses, vegetables, and vegetable oil	49
			5.1.4.3 Meat, fat, milk, and eggs	51

		5.1.4.4	Animal feed	53
		5.1.4.5	Miscellaneous products	53
	5.1.5	Terrestrial and aquatic organisms		54
		5.1.5.1	Plants	54
		5.1.5.2	Aquatic organisms	54
		5.1.5.3	Terrestrial organisms	55
5.2	Exposure of the general population			56
	5.2.1	Total-diet studies		57
	5.2.2	Intake with drinking-water and air		59
	5.2.3	Concentrations in human samples		59
		5.2.3.1	Blood	59
		5.2.3.2	Adipose tissue	61
		5.2.3.3	Breast milk	61

6. KINETICS AND METABOLISM 64

6.1	Absorption			64
	6.1.1	Oral administration—experimental animals		64
	6.1.2	Dermal application—experimental animals		64
	6.1.3	Other routes—experimental animals		65
6.2	Distribution			66
	6.2.1	Oral administration—experimental animals		66
	6.2.2	Inhalation—experimental animals		67
	6.2.3	Other routes		67
6.3	Metabolic transformation			69
	6.3.1	Enzymatic involvement		71
	6.3.2	Identification of metabolites		73
	6.3.3	Metabolites identified in humans		74
6.4	Elimination and excretion in expired air, faeces, and urine			74
	6.4.1	Oral administration		74
		6.4.1.1	Rat	75
		6.4.1.2	Rabbit	75
	6.4.2	Other routes		76
		6.4.2.1	Mouse	76
		6.4.2.2	Rat	76
		6.4.2.3	Human	77
6.5	Retention and turnover (experimental animals)			77
6.6	Biotransformation			78
	6.6.1	Plants		78
	6.6.2	Microorganisms		81
		6.6.2.1	Anaerobic conditions	83
		6.6.2.2	Aerobic conditions	83
6.7	Isomerization			83

7.	EFFECTS ON LABORATORY MAMMALS AND IN *IN-VITRO* TEST SYSTEMS		86
	7.1 Single exposure		86
	7.1.1 Oral		86
	7.1.2 Intraperitoneal and intramuscular		87
	7.1.3 Inhalation		87
	7.1.4 Dermal		88
	7.2 Short-term exposure		89
	7.2.1 Oral		89
	7.2.1.1 Mouse		89
	7.2.1.2 Rat		89
	7.2.1.3 Dog		91
	7.2.1.4 Pig		92
	7.2.2 Inhalation		92
	7.2.2.1 Mouse		92
	7.2.2.2 Rat		93
	7.2.3 Dermal		93
	7.3 Skin and eye irritation; sensitization		94
	7.3.1 Primary skin irritation		94
	7.3.2 Primary eye irritation		94
	7.3.3 Sensitization		94
	7.4 Long-term exposure		95
	7.4.1 Oral		95
	7.4.2 Appraisal of acute and short- and long-term studies		96
	7.5 Reproduction, embryotoxicity, and teratogenicity		96
	7.5.1 Reproduction		96
	7.5.2 Embryotoxicity and teratogenicity		97
	7.5.2.1 Oral administration		97
	7.5.2.2 Subcutaneous injection		98
	7.5.3 Reproductive behaviour		99
	7.5.4 Appraisal of reproductive toxicology		99
	7.6 Mutagenicity and related end-points		100
	7.6.1 DNA damage		100
	7.6.2 Mutation		101
	7.6.3 Chromosomal effects		101
	7.6.4 Miscellaneous tests		107
	7.6.5 Appraisal of mutagenicity and related end-points		107
	7.7 Carcinogenicity		108
	7.7.1 Mouse		108
	7.7.2 Rat		111
	7.7.3 Initiation–promotion		112
	7.7.4 Mode of action		113
	7.7.5 Appraisal of carcinogenicity		114

7.8 Special studies 115
    7.8.1 Immunosuppression 115
    7.8.2 Behavioural studies 115
    7.8.3 Neurotoxicity 115
        7.8.3.1 Dose–response studies using intact animals 115
        7.8.3.2 Studies on mechanism 118
        7.8.3.3 Summary 120
7.9 Factors that modify toxicity; toxicity of metabolites 121

8. EFFECTS ON HUMANS 122

    8.1 Exposure of the general population 122
        8.1.1 Acute toxicity, poisoning incidents 122
        8.1.2 Effects of short- and long-term exposures—
            controlled human studies 123
            8.1.2.1 Oral administration 123
            8.1.2.2 Dermal application 123
        8.1.3 Epidemiological studies (general population) 125
    8.2 Occupational exposure 125
        8.2.1 Toxic effects 125
        8.2.2 Irritation and sensitization 129

9. EFFECTS ON OTHER ORGANISMS IN THE LABORATORY
    AND FIELD 131

    9.1 Microorganisms 131
        9.1.1 Bacteria 131
        9.1.2 Algae 131
            9.1.2.1 Blue-green algae 131
            9.1.2.2 Freshwater algae 132
            9.1.2.3 Marine algae 132
        9.1.3 Dinoflagellates, flagellates, and ciliates 132
    9.2 Aquatic organisms 133
        9.2.1 Invertebrates 133
            9.2.1.1 Crustacea 134
            9.2.1.2 Aquatic arthropods 134
            9.2.1.3 Molluscs 134
        9.2.2 Fish 137
            9.2.2.1 Acute toxicity 137
            9.2.2.2 Short- and long-term toxicity 138
            9.2.2.3 Reproduction 143
        9.2.3 Amphibia 143
            9.2.3.1 Acute toxicity 143
            9.2.3.2 Effects on hatching and larval development 143

	9.3	Terrestrial organisms	145
		9.3.1 Honey-bees	145
		9.3.2 Birds	145
		9.3.2.1 Acute toxicity	145
		9.3.2.2 Short-term toxicity	145
		9.3.2.3 Reproduction	148
		9.3.3 Mammals	148
	9.4	Appraisal	149
10.	PREVIOUS EVALUATIONS BY INTERNATIONAL BODIES		150
APPENDIX I			152
REFERENCES			154
RESUMÉ			191
RESUMEN			200

WHO TASK GROUP MEETING ON ENVIRONMENTAL HEALTH CRITERIA FOR LINDANE

Members

Dr S. Dobson, Pollution and Ecotoxicology Section, Institute of Terrestrial Ecology, Monkswood Experimental Station, Abbots Ripton, Huntingdon, United Kingdom

Dr G.J. van Esch, Bilthoven, The Netherlands (*Joint Rapporteur*)

Dr M. Herbst, Biological Research, ASTA Pharma AG, Frankfurt, Germany (*Joint Rapporteur*)

Professor J.S. Kagan, Department of General Toxicology and Experimental Pathology, All-Union Scientific Research Institute of Hygiene and Toxicology of Pesticides, Polymers, and Plastics, Kiev, USSR (*Vice-Chairman*)

Dr S.G.A. Magwood, Pesticides Division, Environmental Health Centre, Health and Welfare Canada, Tunney's Pasture, Ottawa, Ontario, Canada

Professor W.-O. Phoon, National Institute of Occupational Health and Safety, University of Sydney, Sydney, Australia (*Chairman*)

Dr J.F. Risher, US Environmental Protection Agency, Environmental Criteria and Assessment Office, Cincinnati, Ohio, USA

Dr Y. Saito, Division of Foods, National Institute of Hygienic Sciences, Setagaya-ku, Tokyo, Japan

Dr V. Turusov, Laboratory of Carcinogenic Substances, All-Union Cancer Research Centre, Moscow, USSR

Representatives of Non-Governmental Organizations

    Dr P.G. Pontal, International Group of National Associations of Manufacturers of Agrochemical Products (GIFAP), Brussels, Belgium

Observers

    Dr A.V. Bolotny, All-Union Scientific Research Institute of Hygiene and Toxicology of Pesticides, Polymers, and Plastics, Kiev, USSR

    Dr D. Demozay, International Centre for the Study of Lindane (CIEL), Rhône-Poulenc Agrochimie, Lyon, France

Secretariat

    Dr G.J. Burin, International Programme on Chemical Safety, World Health Organization, Geneva, Switzerland

    Dr K.W. Jager, International Programme on Chemical Safety, World Health Organization, Geneva, Switzerland (*Secretary*)

    Dr V.A. Rezepov, Centre for International Projects, USSR State Committee for Environmental Protection, Moscow, USSR

NOTE TO READERS OF THE CRITERIA DOCUMENTS

Every effort has been made to present information in the criteria documents as accurately as possible without unduly delaying their publication. In the interest of all users of the environmental health criteria documents, readers are kindly requested to communicate any errors that may have occurred to the Manager of the International Programme on Chemical Safety, World Health Organization, Geneva, Switzerland, in order that they may be included in corrigenda, which will appear in subsequent volumes.

* * *

A detailed data profile and a legal file can be obtained from the International Register of Potentially Toxic Chemicals, Palais des Nations, 1211 Geneva 10, Switzerland (Telephone No. 7988400 or 7985850)

* * *

The proprietary information contained in this document cannot replace documentation for registration purposes, because the latter has to be closely linked to the source, the manufacturing route, and the purity/impurities of the substance to be registered. The data should be used in accordance with paragraphs 82–84 and recommendations paragraph 90 of the Second FAO Government Consultation (1982).

ENVIRONMENTAL HEALTH CRITERIA FOR LINDANE

The WHO Task Group on Environmental Health Criteria for Lindane met in Moscow, USSR, on 20–24 November 1989. The meeting was convened with the financial assistance of the United Nations Environment Programme (UNEP) and was hosted by the Centre for International Projects (CIP), USSR State Committee for Environmental Protection. On behalf of the CIP, Dr V.A. Rezepov opened the meeting and welcomed the participants. Dr K.W. Jager welcomed the participants on behalf of the three cooperating organizations of the IPCS (UNEP, ILO, WHO). The Task Group reviewed and revised the draft document and made an evaluation of the risks to human health and the environment from exposure to lindane.

The first drafts of this monograph were prepared by Dr M. Herbst (on behalf of the International Centre for the Study of Lindane (CIEL)) and Dr G.J. van Esch (on behalf of the IPCS). The second draft was prepared by Dr G.J. van Esch, incorporating comments received following circulation of the first draft to the IPCS contact points for Environmental Health Criteria publications.

The help of CIEL in making available its proprietary toxicological information on lindane to IPCS and the Task Group is gratefully acknowledged. This enabled the Task Group to make its evaluation on the basis of more complete data than would otherwise have been possible.

The efforts of all who helped in the preparation and finalization of the document are also gratefully acknowledged. Dr K.W. Jager of the IPCS Central Unit was responsible for the technical development of this monograph and Mrs E. Heseltine of St Léon-sur-Vézère, France, for the editing.

1. SUMMARY AND EVALUATION; CONCLUSIONS; RECOMMENDATIONS

1.1 Summary and evaluation

1.1.1 General properties

Technical-grade hexachlorocyclohexane (HCH) consists of 65–70% alpha-HCH, 7–10% beta-HCH, 14–15% gamma-HCH, and approximately 10% of other isomers and compounds. Lindane contains > 99% gamma-HCH. It is a solid, with a low vapour pressure, and is poorly soluble in water but very soluble in organic solvents, such as acetone, and in aromatic and chlorinated solvents. The n-octanol/water partition coefficient (log P_{ow}) is 3.2–3.7.

Lindane can be determined separately from the other isomers of HCH after extraction by liquid/liquid partition, column chromatography and detection by gas chromatography with electron capture. As these analytical methods are highly sensitive, residues of lindane can be identified at a level of nanograms per kilogram or per litre.

Lindane has been used as a broad-spectrum insecticide since the early 1950s for purposes that include treatment of seeds and soil, application on trees, timber and stored materials, treatment of animals against ectoparasites and in public health.

1.1.2 Environmental transport, distribution and transformation

Lindane is strongly adsorbed on soils that contain a large amount of organic matter; furthermore, it can move downward through the soil with water from rainfall or artificial irrigation. Volatilization appears to be an important route of its dissipation under the high-temperature conditions of tropical regions.

Lindane undergoes rapid degradation (dechlorination) in the presence of ultra-violet irradiation, to form pentachlorocyclohexenes (PCCHs) and tetrachlorocyclohexenes (TCCHs). When lindane undergoes environmental degradation under humid or submerged conditions and in field conditions, its half-time varies from a few days to three years, depending on type of

Summary and evaluation; conclusions; recommendations

soil, climate, depth of application and other factors. In agricultural soils common in Europe, its half-time is 40–70 days. The biodegradation of lindane is much faster in unsterilized than in sterilized soils. Anaerobic conditions are the most favourable for its microbial metabolization. Lindane present in water is degraded mostly by microorganisms in sediments to form the same degradation products.

Limited amounts of lindane and gamma-PCCHs are taken up by and translocated into plants, especially in soils with a high content of organic matter. Residues are found mainly in the roots of plants, and little, if any, is translocated into stems, leaves or fruits. Rapid bioconcentration takes place in microorganisms, invertebrates, fish, birds and humans, but biotransformation and elimination are relatively rapid when exposure is discontinued. In aquatic organisms, uptake from water is more important than uptake from food. The bioconcentration factors in aquatic organisms under laboratory conditions ranged from approximately 10 up to 6000; under field conditions, the bioconcentration factors ranged from 10 to 2600.

1.1.3 Environmental levels and human exposure

Lindane has been found in the air above the oceans at concentrations of 0.039–0.68 ng/m^3 and has been measured at up to 11 ng/m^3 in the air in some countries. The estimated concentrations in surface water in a number of European countries were mainly below 0.1 µg/litre. The concentration in the River Rhine and its tributaries in 1969–74 varied between 0.01 and 0.4 µg/litre; after 1974, it was below 0.1 µg/litre. Levels of 0.001–0.02 µg/litre have been reported in sea water. The concentrations of lindane in soil are generally low—in the range 0.001–0.01 mg/kg, except in areas where waste is disposed of.

Fish and shellfish have been found to contain gamma-HCH at concentrations ranging from none detected up to 2.5 mg/kg on a fat basis, depending on whether they live in fresh or sea water and whether they have a low or high fat content. Levels of about 330 and 440 µg/kg (wet weight) were found in adipose tissue of polar bears in 1982 and 1984, respectively. The concentration of lindane in the livers of birds of prey varied between 0.01 and 0.1 mg/kg. Eggs of sparrow- hawks collected in 1972–73 in the Federal Republic of Germany contained levels of 0.6 up to 11.1 mg/kg (on a fat basis).

The concentrations of lindane in drinking-water are generally below 0.001 µg/litre, and in industrialized countries more than 90% of human intake of lindane originates from food. Over the last 25 years, selected food items have been analysed for lindane in a large number of countries. The concentrations found in cereals, fruits, vegetables, pulses, and vegetable oils ranged from not detected up to 5 mg/kg product, and those in milk, fat, meat, and eggs from not detected up to 5.1 mg/kg product (on a fat basis). In only a few instances were higher concentrations found. The concentrations in fish were generally far lower than 0.05 mg/kg product (on a fat basis). In total-diet and market-basket studies to estimate daily human intake of lindane, a clear difference was observed with time: intake in the period around 1970 was up to 0.05 µg/kg body weight per day, whereas by 1980 intake had decreased to 0.003 µg/kg body weight per day or lower. In the USA, the daily intake of gamma-HCH between 1976 and 1979 decreased from 0.005 to 0.001 µg/kg body weight per day for infants and from 0.01 to 0.006 µg/kg body weight per day for toddlers.

Determinations of the lindane content in body tissues in the general population have been made in a number of countries. The content in blood in the Netherlands was in the order of < 0.1–0.2 µg/litre, but much higher concentrations were found in several countries where technical-grade HCH was used. The mean concentrations in human adipose tissue in various countries ranged from < 0.01 up to 0.2 mg/kg on a fat basis. The concentrations of lindane in human milk are generally rather low, at average concentrations of < 0.001 up to 0.1 mg/kg on a fat basis; however, there has been a clear decrease over time.

Lindane is thus distributed all over the world and can be detected in air, water, soil, sediment, aquatic and terrestrial organisms, and food, although the concentrations in these different compartments are generally low and are gradually decreasing. Humans are exposed daily via food, and lindane has been found in blood, adipose tissue, and breast milk; the levels of intake, however, are also decreasing.

1.1.4 *Kinetics and metabolism*

In rats, lindane is absorbed rapidly from the gastrointestinal tract and distributed to all organs and tissues within a few hours. The highest concentrations are found in adipose tissues and skin; in various studies, the fat:blood ratio was about 150–200, the liver:blood ratio, 5.3–9.6 and the brain:blood ratio, 4–6.5. The same fat:blood ratio was found in rats exposed by inhalation. These ratios vary with sex, being higher in females. Uptake of lindane through the skin after dermal application is slow and occurs to a very limited extent; this may explain the low toxicity of lindane after dermal exposure.

Summary and evaluation; conclusions; recommendations

Lindane is metabolized mainly in the liver by four enzymatic reactions: dehydrogenation to gamma-HCH, dehydrochlorination to gamma-PCCH, dechlorination to gamma-TCCH and hydroxylation to hexachlorocyclohexanol. The end-products of biotransformation are di-, tri-, tetra-, penta-, and hexachloro- compounds. These metabolites are excreted mainly via the urine in the free form or conjugated with glucuronic acid, sulfuric acid or N-acetylcystein. The elimination is relatively fast, with half-times in rats of 3–4 days. Bacteria and fungi metabolize lindane to TCCH and PCCH. The rate of metabolic transformation in plants is low, and the main degradation pathway proceeds via PCCH to tri- and tetrachlorophenol and conjugates with beta-glucose and other, unknown compounds. There is no evidence that lindane is isomerized to alpha-HCH.

1.1.5 Effects on organisms in the environment

Lindane is not very toxic for bacteria, algae, or protozoa: 1 mg/litre was generally the no-observed effect level (NOEL). Its action on fungi is variable, with NOELs varying from 1 to 30 mg/litre depending on the species. It is moderately toxic for invertebrates and fish, the $L(E)C_{50}$ values for these organisms being 20-90 µg/litre. In short-term and long-term studies with three species of fish, the NOEL was 9 µg/litre; no effect on reproduction was seen with levels of 2.1–23.4 µg/litre. The LC_{50} values for both freshwater and marine crustacea varied between 1 and 1100 µg/litre. Reproduction in *Daphnia magna* was depressed in a dose-dependent fashion; the NOEL was in the range 11–19 µg/litre. Reproduction of molluscs was not adversely effected by a dose of 1 mg/litre.

The LD_{50} for honey-bees was 0.56 µg/bee.

Acute oral LD_{50} values for a number of bird species were between 100 and 1000 mg/kg body weight. In short-term studies with birds, doses of 4–10 mg/kg diet had no effect, even on egg-shell quality. Laying ducks treated with doses of lindane up to 20 mg/kg body weight, however, had decreased egg production.

Bats exposed to wood shavings that initially contained 10–866 mg/m^2 lindane, resulting from application at the recommended rate, all died within 17 days. No data were available on effects on populations and ecosystems.

1.1.6 Effects on experimental animals and in vitro

The acute oral toxicity of lindane is moderate: the LD_{50} for mice and rats is in the range 60–250 mg/kg body weight, depending on the vehicle used. The dermal LD_{50} for rats is approximately 900 mg/kg body weight. Toxicity was manifested by signs of central nervous system stimulation.

Lindane does not irritate or sensitize the skin; it is slightly irritating to the eye.

In a 90-day study in rats, the NOEL was 10 mg/kg diet (equivalent to 0.5 mg/kg body weight). At 50 and 250 mg/kg diet, the weights of the liver, kidneys, and thyroid were increased; at 250 mg/kg diet, an increase was seen in liver enzyme activity. This increase in enzyme activity accelerates the breakdown of both lindane and other compounds. In another 90-day study in rats, 4 mg/kg diet (equivalent to 0.2 mg/kg body weight) was considered to be the no-observed-adverse-effect level (NOAEL); renal and hepatic toxicity were observed at concentrations of 20 mg/kg diet and higher. A short-term toxicity study in mice was considered to be inadequate to establish a NOEL.

Administration of lindane to dogs at 15 mg/kg in the diet (equivalent to 0.6 mg/kg body weight) for 63 weeks had no toxic effect. In a two-year study of the toxicity of this compound in dogs, in which a large number of parameters were measured, no treatment-related abnormality was apparent at doses of 50 mg/kg diet (equivalent to 2 mg/kg body weight) and lower. In the group given 100 mg/kg diet, however, levels of alkaline phosphatase were increased; and with 200 mg/kg diet, abnormalities in electro-encephalogram tracings indicative of non-specific neuronal irritation were observed.

In rats exposed by inhalation to lindane at 0.02–4.54 mg/m^3 for 6 h/day for 3 months, the highest dose induced increases in hepatic cytochrome P450 values; the NOAEL was found to be 0.6 mg/m^3. In two long-term studies in rats, carried out many years ago, doses of 10–1600 mg/kg diet were tested. In one of these studies, 50 mg/kg diet (equivalent to 2.5 mg/kg body weight) was found to be the NOAEL. At 100 mg/kg diet, an increase in liver weight, hepatocellular hypertrophy, fatty degeneration, and necrosis were found. In the other study, 25 mg/kg diet (equivalent to 1.25 mg/kg body weight) had no effect, but hepatocellular hypertrophy and fatty degeneration were seen with 50 mg/kg diet.

Lindane has been investigated for its effects on all aspects of reproduction (in rats over three generations) and for its embryotoxicity and teratogenicity after oral, subcutaneous and intraperitoneal administration in mice, rats, dogs, and pigs. It had no teratogenic effect after oral or parenteral administration (extra ribs were regarded as variations). Fetotoxic and/or maternal toxic effects were observed with doses of 10 mg/kg body weight and above given by oral gavage; 5 mg/kg body weight is considered to be the NOAEL. Lindane had no effect on reproduction or maturation in the three-generation study in rats at doses of up to 100 mg/kg diet; but with 50 mg/kg diet, morphological changes in the liver indicating enzyme induction occurred in the offspring of the third generation. The NOEL in this test was 25 mg/kg diet (equivalent to 1.25 mg/kg body weight).

The NOEL for neurotoxicity in a 22-day study in rats was 2.5 mg/kg body weight.

The mutagenicity of lindane has been studied adequately. In extensive investigations of its ability to induce gene mutations in bacteria and in mammalian cells, and for its capacity to induce sex-linked recessive lethal mutations in *Drosophila melanogaster*, negative results were obtained consistently. Lindane also gave negative results in tests for chromosomal damage and sister chromatid exchange in mammalian cells *in vitro* and *in vivo*. The results of assays for DNA damage in bacteria and for covalent binding to DNA in the liver of rats and mice *in vivo* following oral administration were also negative. In the very few studies in which positive results were obtained, either the study design was invalid or the purity of the compound tested was not reported. Overall, however, lindane appears to have no mutagenic potential.

Studies to define the carcinogenic potential of lindane have been carried out in mice and rats using dose levels of up to 600 mg/kg diet in mice and up to 1600 mg/kg diet in rats. Hyperplastic nodules and/or hepatocellular adenomas were found in mice given doses of 160 mg/kg diet or more; in some studies, the dose levels exceeded the maximum tolerated dose. Two studies in mice with dose levels of up to 160 mg/kg diet and one in rats with 640 mg/kg diet showed no increase in the incidence of tumours.

The results of studies on initiation-promotion of carcinogenicity, on the mode of action, and on mutagenicity indicate that the tumorigenic response observed with gamma-HCH in mice is mediated by a nongenetic mechanism.

1.1.7 Effects on humans

Several cases of fatal poisoning and of non-fatal illness caused by lindane have been reported, which were either accidental, intentional (suicide), or due to gross neglect of safety precautions or improper uses of medical products containing lindane. Symptoms included nausea, restlessness, headache, vomiting, tremor, ataxia, tonic–clonic convulsions and/or changes in the electroencephalographic pattern. These effects were reversible after discontinuation of exposure or symptomatic treatment.

Notwithstanding extensive use over 40 years, very few cases of poisoning in the occupational setting have been reported. In workers exposed for long periods during either manufacture or application of lindane, the only sign found was increased activity of drug-metabolizing enzymes in the liver. There is no evidence for the relationship suggested in some publications between exposure to lindane and the occurrence of blood dyscrasias. A few acute and short-term studies in humans indicate that a dose of approximately 1.0 mg/kg body weight does not induce poisoning; however, a dose of 15–17 mg/kg body weight resulted in severe toxic symptoms.

Approximately 10% of a dermally applied dose is absorbed, although more passes through damaged skin.

1.2 Conclusions

1.2.1 General population

Lindane is circulating in the environment and is present in food chains, so that humans will continue to be exposed. The daily intake and total exposure of the general population are decreasing gradually; however, they are clearly below the advised acceptable daily intake and are of no concern to public health.

1.2.2 Subpopulations at special risk

The presence of lindane in breast milk results in exposure of breast-fed babies to levels that are generally below the acceptable daily intake and therefore of no concern to health. Although lower levels of exposure would be preferred, the present levels are not a limiting factor for the practice of natural breast-feeding.

Summary and evaluation; conclusions; recommendations

Prescriptions should be followed strictly with regard to the therapeutic use of lindane against scabies and to control body lice.

1.2.3 Occupational exposure

As long as the recommended precautions to minimize exposure are observed, lindane can be handled safely.

1.2.4 Environmental effects

Lindane is toxic to bats that roost in close contact with wood treated according to the recommended conditions of application. Apart from the results of studies of spills into the aquatic environment, there is no evidence to suggest that the presence of lindane in the environment poses a significant hazard to populations of other organisms.

1.3 Recommendations

1. In order to minimize environmental pollution by other isomers of HCH, lindane (> 99% gamma-HCH) must be used instead of technical-grade HCH.

2. In order to avoid environmental pollution, by-products of and effluents from the manufacture of lindane should be disposed of in an appropriate way.

3. In disposing of lindane, care should be taken to avoid contamination of natural waters and soil.

4. As for other pesticides, proper instructions about application procedures and safety precautions should be given to people who handle lindane.

5. Long-term carcinogenicity tests conducted according to present-day standards should be conducted.

6. Monitoring of the daily intake of lindane by the general population should continue.

2. IDENTITY, PHYSICAL AND CHEMICAL PROPERTIES, ANALYTICAL METHODS

2.1 Identity

2.1.1 Primary constituent

Common name: Lindane

Chemical structure[1]:

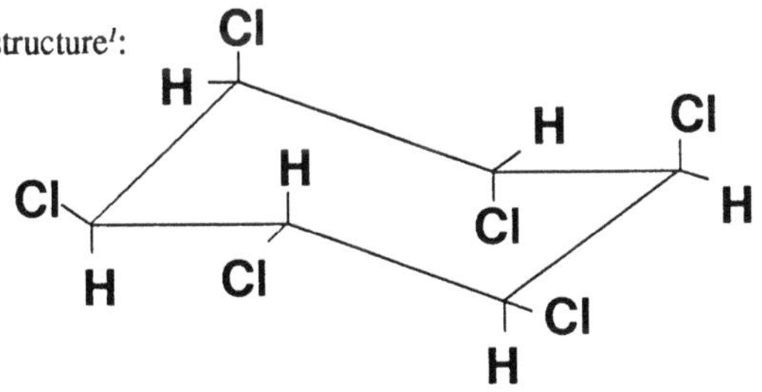

Fig. 1. Chemical structure of lindane

Chemical formula: $C_6H_6Cl_6$

Relative molecular mass: 290.8 (290.9)

CAS chemical name: 1α,2α,3β,5α,6β-hexachlorocyclohexane

CAS registry number: 58-89-9

RTECS registry number: GV4900000

Synonym: Hexachlorocyclohexane (gamma-isomer)

According to IUPAC rules, the designation 'benzene hexachloride' is incorrect; nevertheless, it is still widely used, especially in the form of its abbreviation, BHC. This is therefore another common name approved by the ISO. The compound is called gamma-HCH by WHO, but gamma-BHC by FAO (FAO, 1973). The synonym hexachlorocyclohexane (gamma

---
[1]See Appendix I

Identity, physical and chemical properties, analytical methods

isomer) is used by the Environmental Protection Agency and the American Conference of Governmental Industrial Hygienists in the USA. The definitions of these different appellations are given in Table 1.

Table 1. Definitions of appellations of lindane

Name	Definition	Remarks
Lindane	product containing not less than 99% gamma-HCH	ISO-AFNOR name for a product (not yet recognized by BSI)
Lindane	= gamma-HCH	Common name used for gamma-HCH in the USSR only
γ-HCH	gamma isomer of 1,2,3,4,5,6-hexachlorocyclohexane	ISO-AFNOR common name
γ-BHC	gamma isomer of 1,2,3,4,5,6-benzene hexachloride	ISO BSI common name in English-speaking countries (recognized by ISO as synonym of gamma-HCH)

2.1.2 Technical product

Common trade names: A great number of products containing lindane are on the market; no attempt has been made to list the hundreds of trade names here (see Hudson et al., 1984; Hill & Camardese, 1986; International Register for Potentially Toxic Chemicals, 1989).

Purity: FAO (1973) requires that lindane "... shall consist, essentially, of γ-BHC as white or nearly white granules, flakes or powder, free from extraneous impurities or added modifying agents and with not more than a faint odour." FAO further requires that it contain not less than 99.0% gamma-HCH and that the melting-point be at least 112 °C, which is not depressed when the sample is mixed with an equal amount of pure gamma-HCH.

In some processes for manufacturing lindane, low levels of dioxin may be formed (US Environmental Protection Agency, 1985). Under appropriate manufacturing conditions, however, no 2,3,7,8-tetrachlorodibenzodioxin or 2,3,7,8-tetrachlorodibenzofuran is detected in HCH, lindane, trichlorobenzene, industrial liquid or gaseous effluents at the analytical limit of detection of 1 µg/kg (letter from D. Demozay, Rhône-Poulenc, to IPCS dated 17 November 1989).

2.2 Physical and chemical properties

Lindane is a colourless, crystalline solid with either a faint or no smell (the characteristic smell of technical-grade HCH is attributed to impurities, particularly heptachlorocyclohexane).

Melting-point:	112.8 °C
Boiling-point:	288 °C
Vapour pressure:	0.434 x 10^{-5} kPa (3.26 x 10^{-5} mmHg) at 20 °C 60.6 x 10^{-5} kPa (45.6 x 10^{-5} mmHg) at 40 °C
Density:	1.85
Solubility:	nearly insoluble in water at 20 °C (10 mg/litre); moderately soluble in ethanol (6.7%); slightly soluble in mineral oils; soluble in acetone and in aromatic and chlorinated solvents
Stability:	stable to light, air, heat, carbon dioxide, and strong acids; dehydrochlorinates in the presence of alkali or on prolonged exposure to heat with the formation of trichlorobenzenes, phosgene, and hydrochloric acid. It is incompatible with strong bases and powdered metals, such as iron, zinc, and aluminium, and with oxidizing agents; can undergo oxidation when in contact with ozone.
Corrosivity:	corrosive to aluminium
Inflammability:	not inflammable
n-Octanol/water partition coefficient (log P_{ow}):	3.2–3.7 (see section 4.2.7.1) (Demozay & Marechal, 1972; Dutch Chemical Industry Association, 1980; American Conference of Governmental Industrial Hygienists, 1986; Rhône-Poulenc Agrochimie, 1986)

2.3 Conversion factors

1 ppm = 12.1 mg/m³
1 mg/m³ = 0.083 ppm

2.4 Analytical methods

2.4.1 Sampling

Sampling procedures and methods for preparing samples of formulations and for analysing residues have been described by Mestres (1974), the Deutsche Forschungsgemeinschaft (1979), the Association of Official Analytical Chemists (1980), and Hildebrandt et al. (1986).

2.4.2 Analytical methods

Products are analysed by a cryoscopic method (Raw, 1970; FAO, 1973; WHO, 1985). Formulated products can be analysed by determining hydrolysable chlorine (Raw, 1970; FAO, 1973). Since the latter method is not specific, other methods, such as gas chromatography, are used to obtain sufficient separation of the HCH isomers.

Residues in food and in soil can be determined after adequate clean-up by gas chromatography and other chromatographic methods (Nash et al., 1973; Eichler, 1977; Association of Official Analytical Chemists, 1980; Deutsche Forschungsgemeinschaft, 1983). The principle of the method is extraction of a sample with organic solvents (acetonitrile, hexane/acetone, acetone, and others). Fat is extracted from fatty foods and partitioned between petroleum ether and acetonitrile by extracting aliquots or an entire solution of acetonitrile into petroleum ether. Residues are purified by chromatography on a Florisil colum, and eluted with a mixture of petroleum ether and ethylether. Concentrated residues are measured by gas chromatography with electron capture detection.

The method described by the Deutsche Forschungsgemeinschaft (1979) for fruits and vegetables is based on extraction of samples with acetone and extraction of the aliquot with dichloromethane. The residue obtained after evaporation of the solvent is cleaned by co-distillation, and the distillate is analysed by gas chromatography with electron capture detection. The limit of determination depends on the method, the substrate and the sample size; the lower limit of determination is 0.001–0.01 mg/kg.

Palmer & Kolmodin-Hedman (1972) analysed air samples by gas chromatography with electron capture detection, and alpha-, beta-, and gamma-HCH were determined in serum by gas chromatography after a deproteinization extraction step (Palmer & Kolmodin-Hedman, 1972; Angerer & Barchet, 1983).

Wittlinger & Ballschmiter (1987) provided an extensive description of analytical methods for HCHs in air, involving sampling by adsorption, extraction and preseparation, and determination by high-resolution gas chromatography. Sampling was performed by pumping air through a glass-fibre filter and then through a silica-gel layer, using an internal standard. The sample was extracted with dichloromethane and the extract evaporated. The preseparation was done on silica gel, and the aliquot was eluted in a mixture of hexane and dichloromethane. High-resolution capillary gas chromatography, electron capture detection and a mass selective detector were used for determination.

Eder et al. (1987) described three analytical methods for the determination of HCHs in sediments: moist samples are extracted with a solvent or a mixture of solvents, concentrated or fractionated and determined by gas chromatography and electron capture detection.

Greve (1972) described a method for the determination of organo-chlorine pesticides in water based on gas chromatography of a petroleum ether extract after clean-up over Florisil or silica gel. The limit of detection for lindane is 0.01 µg/litre.

Methods used for the determination of lindane in samples of soil, animal, and vegetable products in the USSR are described by Izmerov (1983). These methods are based on extraction with organic solvents, purification and concentration of the extracts and determination by gas-liquid chromatography with electron-capture detection.

3. SOURCES OF HUMAN AND ENVIRONMENTAL EXPOSURE

3.1 Natural occurrence

Lindane is not known to occur as a natural product.

3.2 Man-made sources

3.2.1 Production levels and processes

3.2.1.1 Manufacturing process

HCH was discovered in 1825, but its insecticidal properties were first patented only in the 1940s. It has been produced commercially since 1949.

Technical-grade HCH is synthesized from benzene and chlorine in the presence of ultra-violet light and comprises 65–70% alpha-HCH, 7–10% beta-HCH, 14–15% lindane (gamma-HCH), approximately 7% delta-HCH, 1–2% epsilon-HCH, and 1–2% other components. By-products can be minimized by careful control of the reaction conditions. Lindane (> 99% gamma-HCH) can be purified by multiple extractions with methanol.

The extraction of lindane from HCH produces 85% non-insecticidal HCH isomers, which can be used as intermediates in the production of trichlorobenzene and hydrochloric acid after cracking in an integrated installation. Trichlorobenzene is used in the synthesis of other chemicals (van Velsen, 1986; Rhône-Poulenc Agrochimie, 1986).

3.2.1.2 World-wide production figures

Lindane is produced in Austria, France, and Spain and in China, India, Turkey, and the USSR. Before 1984, lindane was also manufactured in the German Democratic Republic, Poland, Yugoslavia, Romania, and Hungary; since then, all production has been stopped in Germany, Japan, the Netherlands, the United Kingdom, and the USA.

Although in most developed countries use of technical-grade HCH has been prohibited, it is still used elsewhere on a large scale: total consumption of technical-grade HCH in India in 1986–87 was approximately 27 000 tonnes (International Atomic Energy Agency, 1988).

3.2.2 Emissions

According to De Bruijn (1979), approximately 0.1% of the lindane processed reaches the waste-water of a formulating plant. Treatment of the waste-water, however, leads to solid waste, which should be incinerated. In the past, it was often dumped in the environment and could be dispersed from (open) chemical dumping grounds to more remote soils by the wind.

Lindane enters the environment following application of lindane-containing pesticides. Emissions can cross national boundaries in water and air. For instance, the total trans-frontier flux of lindane into the Netherlands via the surface water of the River Rhine was approximately 1.8 tonnes per year (average for 1980–83) and that via the River Meuse, 0.2 tonnes per year (Slooff & Matthijsen, 1988).

3.2.3 Uses

Lindane is a broad-spectrum insecticide, which has been used since 1949 for agricultural as well as non-agricultural purposes. Approximately 80% of the total production is used in agriculture (Demozay & Marechal, 1972), mostly for seed and soil treatment. Wood and timber protection is the major non-agricultural use. Lindane is also used against ectoparasites in veterinary and pharmaceutical products (Rhône-Poulenc Agrochimie, 1986).

3.2.4 Extent of use

Lindane is used worldwide, with the major exception of Japan, where all uses of HCH were cancelled in 1971 mainly because of environmental pollution with alpha- and beta-HCH resulting from extensive use of technical-grade HCH. At that time, no clear difference was made between the risks presented by the individual HCH isomers, and lindane was banned as well. In almost all other countries, lindane is registered for one or more applications, although the use pattern differs from one country to another.

In 1979, the US Department of Agriculture and the Environmental Protection Agency summarized the percentage uses of lindane in the USA as follows: seed treatment 48%, hardwood lumber 23%, livestock 16%, pets 3%, pecans 3%, pineapples 2%, ornamentals 2%, household 1%, cucurbits 1%, forestry 0.5%, and structures 1%. In France and Germany, 70–80% of all lindane used agriculturally is for soil treatment, to protect maize and

Sources of human and environmental exposure

sugar beets, and 15–20% is used for seed treatment. De Bruijn (1979) reported an estimate of the pattern of use of lindane in the European Economic Community.

3.2.5 Formulations

Formulation facilities exist in many countries. Lindane is made in numerous forms, the most important of which are: wettable powders (up to 90% active ingredient); emulsifiable concentrates (not more than 20% active ingredient); flowable suspensions (in water); solutions in organic solvents (up to 50% active ingredient); dusts and powders (0.5–2% active ingredient); granules and coarse dusts (3–4% active ingredient); ready-for-use baits; aerosols; and special formulations for use in human and veterinary medicine (Demozay & Marechal, 1972).

Lindane dissolved in organic solvents may be used in 'thermal foggers' in glasshouses or atomized in open areas; such solutions are appropriate for aerial application (5–10 litres/ha of formulations containing 5–10% active ingredient). Concentrated solutions containing an anti-vaporization component have been applied using an ultra-low volume method at 0.5–1 litre/ha. Various fumigation preparations for indoor use have been sold, including fumigation strips, tablets, and smoke generators. These contained virtually pure lindane to which a small quantity of binding material was added. Because of its versatility and relatively low acute toxicity, lindane is often used in mixed formulations with other insecticides and fungicides (Demozay & Marechal, 1972).

4. ENVIRONMENTAL TRANSPORT, DISTRIBUTION, AND TRANSFORMATION

4.1 Transport and distribution between media

4.1.1 Volatilization

Some of the active ingredient of lindane volatilizes after it has been applied to control insect pests, especially on leaves. Starr & Johnson (1968) demonstrated that 20% of an applied dose had evaporated 96 h after bean plants had been sprayed with lindane at 16 °C. The evaporation was dependent on temperature and on the humidity of the air.

Some of the lindane that reaches the soil may also vaporize as degradation products. Cliath & Spencer (1972) showed the presence of vapours of the metabolite PCCH, which has a vapour pressure approximately 14 times higher than that of lindane.

In a model test, four soil types, ranging from a loamy sand to a clay, were treated with ^{14}C-lindane to give a concentration of 10 mg/kg soil; water was then added to the samples, they were air-dried at 33 °C and at 55 °C, and volatilization was measured by trapping the vapours. Three cycles of about 14 days each were followed. Lindane volatilized from the soils only with water, and no further volatilization occurred after the soils had reached the dry state. The four soil types were associated with different volatilization rates: the highest occurred in loamy sand. In the analysis of vaporized material, unchanged lindane and its degradation products were not differentiated; however, considerable degradation of lindane was found in the soils, and PCCH was identified as a metabolite. At least some of the vaporized material may therefore have consisted of degradation products (Guenzi & Beard, 1970).

4.1.2 Precipitation

Evaporation and adsorption to solid particles are important processes in the distribution of lindane. Reverse processes such as deposition from the air and remobilization from silt and sediment also play a part. Buscher et al. (1964) demonstrated that aeration of aqueous solutions of lindane resulted in a loss of 10% over three days, which was ascribed to a co-distillation process as it was greater than could be explained by evaporation alone.

Environmental transport, distribution, and transformation

MacKay & Walkoff (1973) confirmed that evaporation is an important process in the loss of HCH. Lichtenstein & Schulz (1970) found that 16.5% HCH was lost from a non-aerated aqueous solution in 24 h at 30 °C.

The amount of lindane that is distributed by dry deposition depends on the nature of the surface above which the organic components are present. The half-time for dry deposition of HCHs (height of mixing layer, 1000 m) in the Netherlands was calculated to be 2–8 days. On this basis, a rough estimate of the annual flux to soil and water in that country would be 0.5–1.5 tonnes from an outdoor air concentration of 0.4 ng/m^3 (Slooff & Matthijsen, 1988).

4.1.3 Movement in soils

Movement of a substance through the soil profile depends on its adsorption–desorption characteristics in soil/water systems and, to some extent, on its volatility in the soil pores and its diffusion. The adsorption–desorption characteristics of lindane have been the object of a number of studies (Kay & Elrick, 1967; Mills & Biggar, 1969; Baluja et al., 1975; Portmann, 1979; Wahid & Sethunathan, 1979, 1980; Wirth, 1985), all of which showed that lindane is strongly adsorbed to organic soil material and weakly adsorbed to inorganic matter. In the absence of organic matter, the clay content and free iron oxide are implicated in the sorption of lindane (Wahid & Sethunathan, 1979). It can be concluded that the mobility of lindane is very low in soils with a high content of organic matter but might be higher in soils containing little organic matter.

No consensus has been reached in the literature about the possibility that lindane can be remobilized by desorption from polluted soil. Generally, HCH isomers are strongly adsorbed. Under certain conditions— high concentrations of lindane in highly permeable soils with a low organic carbon content (< 0.1%)—a small percentage of the compound may be washed out and reach the groundwater. Nevertheless, the low rate of transport of lindane makes the probability that it will reach groundwater low or very low (Slooff & Matthijsen, 1988).

The diffusion of lindane through soil was investigated by Ehlers et al. (1969a,b) and by Shearer et al. (1973). Diffusion was strongly influenced by the water content of the soil, by the bulk density and by temperature. The diffusion coefficient is nearly zero in soil containing 1% water, but with a water content of 3%, lindane is displaced from the adsorbing surface so that the diffusion coefficient becomes maximal; a further increase in water content reduces the diffusion coefficient. The diffusion of lindane in soils can thus vary between a 'vapour' and a 'non-vapour' phase, depending on the concentration of lindane, the length of time and the water content of the soil.

Leaching of three formulations of lindane was investigated in a series of model studies by Heupt (1974) in different soil types. The test system consisted of 30-cm columns filled with soil to which lindane formulations were applied at application rates corresponding to 6 kg of active ingredient per hectare. Rainfall was simulated at a rate of 200 mm within two days. No lindane was found in the eluate at the limit of detection of 1 µg/litre. In field tests by Cliath & Spencer (1971, 1972), lindane was worked into topsoil of two plots of sandy loam and two of silty clay to a depth of 0–7.5 or 7.5–15 cm, corresponding to an application rate of 21 kg/ha. One of the plots of each soil type received additional irrigation. Almost no movement of lindane was found in the dry plots at the end of the two-year observation period. In the irrigated plots, a broadening of the lindane-containing zone and downward movement to a depth of 60 cm were observed, especially in sandy loam; in clay soil, lindane had almost no mobility.

In a series of dissipation studies with ^{14}C-labelled lindane in soils, coordinated by FAO and the International Atomic Energy Agency, it was demonstrated that persistent pesticides such as lindane dissipate much faster in the tropics than in temperate climates, probably owing to a large extent to volatilization (International Atomic Energy Agency, 1988), as had been found by Edwards (1973a,b, 1977). Table 2 summarizes the results of these studies.

Table 2. Field half-times of gamma-HCH in soils (0–10-cm depth)[a]

Country	Half-time (days)[b]			Time required for initial loss of 50% of radioactivity (days)
	Overall	First phase	Second phase	
India	138 (124–147)	41 (30–50)	188 (83–362)	30–45
Ecuador	150–171	54–60	120–160	40–50
Kenya	5–8	–	–	3–4

[a] Adapted from International Atomic Energy Agency (1988)

[b] In temperate soils, the mean half-time was 438 (401-1022) days (Edwards, 1973a, b)

Environmental transport, distribution, and transformation

4.1.4 Uptake and translocation in plants

One of the first investigations on the absorption of lindane by various seeds was reported by Bradbury & Whitaker (1956). Lindane was taken up from a nutrient solution by the roots of wheat seedlings at a rate of up to 100 mg/kg (fresh weight) within seven days. Subsequent investigations demonstrated that uptake by plants is dependent on a variety of factors.

The influence of *soil type* was investigated by Bradbury (1963). Seedlings grown from seed dressed with lindane and planted in sand had residue levels about two-fold higher than those of plants grown in compost. A further study was reported of the fate of ^{14}C-lindane in loam and sandy soil and in oat plants grown in these soils for 13 days. The loam soil was treated with about 7.3 mg/kg active ingredient and the sandy soil with about 3 mg/kg. Residues were found to be more persistent in loam (53.5% of radiolabel) than in sandy soil (33.8%), but oats grown in sandy soil took up more residues than those grown in loam (loam soil: roots, 0.5%; tops, 0.3%; sandy soil: roots, 2.5%; tops, 1.2%). ^{14}C-Lindane was the major constituent of the soil residues soluble in organic solvents. A major metabolite, which was probably gamma-PCCH, represented 11% of the organic-soluble radiolabelled residues in loam soil; 2,4,5-trichlorophenol accounted for 2.7% of these residues. The authors concluded that the three major factors that determine the environmental fate of ^{14}C-lindane and other insecticides are the insecticide itself, its solubility in water and the type of soil to which it is applied. Compounds with greater aqueous solubility are more mobile, are taken up by plants to a greater extent, and appear to be more susceptible to degradation than compounds less soluble in water. In soils with little organic matter, insecticide residues are more mobile and hence more susceptible to volatilization, uptake by plants, and degradation than in more adsorbent soils such as loam (Fuhremann & Lichtenstein, 1980).

The *rate of application* to soil was found to be a further important factor in determining residue levels. Transfer of lindane from soil into rice plants was almost proportional to the level of contamination of the soil (Kawahara, 1972), but only at low levels of contamination. Charnetski & Lichtenstein (1973) reported a good correlation between the content of ^{14}C-lindane in sand (at up to 6 mg/kg, which is about 12 times the concentration expected after normal application) and that in pea plants grown for six days; at concentrations greater than 10 mg/kg of soil, there was no further increase in the residue levels.

Uptake of lindane after *application to leaves* is lower than that resulting from application to soil. In lettuce and endives treated with ^{14}C-lindane and grown for 21 and 37 days, respectively, only 4.5–13.9% of the applied radioactivity was found at the time of harvest, and most of the lindane had evaporated into the atmosphere (Kohli et al., 1976a).

Differences in residue levels are also dependent on the *plant species*. Of a series of edible crops grown in soil containing lindane at an initial concentration of 5 mg/kg (about 10 times the normal application rate), carrots had higher levels than beans, tomatoes, or potatoes (San Antonio, 1959). More lindane was absorbed from soil with an initial concentration of 2.6 mg/kg by radishes, turnips, and spinach than by Chinese cabbage (Kawahara et al., 1971). The amounts of residues of HCH isomers in turnips were proportional to the initial concentrations of the isomers in the soil (0.05, 0.1, 0.5, 1, 5, 10, or 50 mg/kg soil). The soil:plant residue ratios were in the range 10–20:1 (Kawahara & Nakamura, 1972).

The translocation of lindane and its metabolites in plants has also been investigated in detail (San Antonio, 1959; Bradbury, 1963; Itokawa et al., 1970; Kawahara, 1972; Kawahara & Nakamura, 1972; Charnetski & Lichtenstein, 1973; Balba & Saha, 1974; Eichler, 1980; Korte, 1980). Neither lindane taken up from soil nor its metabolites were evenly distributed within the plants. Comparatively high residue levels were always detected in the roots, whereas only small amounts were translocated into stems, leaves, and fruits. Paasivirta et al. (1988) showed that in water-plants, lindane concentrations are similar in roots and leaves.

1.2 Biotransformation

1.2.1 Degradation

1.2.1.1 Degradation under humid conditions

The half-times of lindane found by different investigators vary considerably, depending on the type of soil to which it is applied and, possibly, temperature. Lindane incubated in a sandy–loam soil with a water capacity of 28% and 60% saturation at room temperature had a half-time of approximately 40 days (Heeschen et al., 1980). The half-times of lindane in model tests were 4–6 weeks in humid sand with a high content of organic matter and 30 weeks in sandy loam (Heupt, 1979). The half-times in aerobic and anaerobic conditions ranged from 12 to 174 days and 100 to 720 days, respectively; in aerobic field conditions, the half-time was 88–1146 days (Edwards, 1966; Kohnen et al., 1975; Kampe, 1980; Rao & Davidson, 1982; MacRae et al., 1984).

Environmental transport, distribution, and transformation

Assuming that lindane is not washed out below the level of ploughed furrows (approximately 20 cm), a half-life of 350 days will result in persistence of 50% of a dose one year after application (Slooff & Matthijsen, 1988). One month after double treatment of potato, beet, and maize crops with lindane, the gamma-HCH content in sandy loam soil was 0.32 mg/kg in the field occupied by maize and 0.68–0.70 mg/kg in the fields with potatoes and beet. After nine months, the lindane content in the beet fields had decreased 14 times and that in the maize fields by only 1.3 times (Kovaleva & Talanov, 1973; see Izmerov, 1983).

4.2.1.2 Degradation under submerged conditions

Half-time values for lindane of a few to about 120 h were determined after incubation in various submerged soil samples. More rapid degradation occurred in soils with a high amino acid content, and the rate also clearly depended on the number of degrading microorganisms present (Ohisa & Yamaguchi, 1979). The rapidity with which lindane was degraded under flooded conditions varied in soil samples from different locations in Japan. Enrichment of the soils with peptone and exclusion of oxygen increased the degradation rate (Ohisa & Yamaguchi, 1978a).

Half-time values of 10–30 days were observed in a comparison of four Philippine rice soils under flooded conditions at a temperature of 30 °C. Lindane was degraded faster at higher temperatures (Yoshida & Castro, 1970). In a similar study with five Indian rice soils at 28 °C, ^{14}C-labelled lindane was degraded at half-times of between 10 days and more than 41 days. Addition of rice straw enhanced the degradation (Siddaramappa & Sethunathan, 1975).

Tsukano (1973) found a half-time for lindane of 10–14 days in soil samples mixed with water. The degradation was almost completely inhibited after addition of sodium azide to the soils, indicating that the degradation observed in non-sterilized soils was due to microbial activity.

4.2.2 Degradation under field conditions

Nash (1983) used a microagroecosystem in which moist fallow sandy loam was placed in a glass chamber at a depth of 15 cm, plants were grown in the chamber and lindane was applied to the surface. A half-time of 1–4 days was found for dissipation of lindane in the soil.

In April 1954, formulations containing lindane were applied to a sandy loam soil at rates of 2.25 and 4.5 kg/ha on field plots in the Rhine valley, and loss of active ingredient was followed during the subsequent 1.5 years using a biological test method. The insecticidal activity disappeared rapidly during the following vegetation period but remained almost constant in winter; further degradation was observed during the second vegetation period. At the end of the observation period, 3.5–5.5% of the lindane applied at 2.25 kg/ha remained, and 17–19.5% of that applied at 4.5 kg/ha: the speed of degradation was therefore greater at the lower application rate. Degradation was virtually identical when the lindane was worked into the soil to a depth of 1–2 cm and when it was introduced to a depth of 10 cm (Schmitt, 1956).

In a field test in Miami, Florida, USA, on silt loam and muck soils, lindane was applied at the extremely high rates of 11.2 or 112.1 kg/ha. The initial half-time at the lower rate was 15.5 months in muck soil and 4.75 months in loam soil. Degradation was slower at the higher rate: the initial half-times were 28.8 months in muck soil and 11.1 months in loam soil (Lichtenstein & Schulz, 1958a). In an earlier study on the same field plots with the same application rates, however, Lichtenstein & Schulz (1958b) found that most of the material detected chemically was inactive in the bioassay and therefore did not represent lindane. They concluded that the breakdown of lindane is faster than it appeared to be using their analytical method.

In an extensive study, sandy loam, silt loam, and muck soils on plots in three midwestern states of the USA were treated with lindane in 1954 at application rates of 1.1, 11.2, and 112.1 kg/ha to a depth of 15.2 cm. After a 4.5-year follow-up, no lindane was detected on plots treated with 1.1 kg/ha; but after application at the higher rates (far in excess of normal rates), about 36% of the applied dose remained. Two major factors that affect the persistence of lindane in soils appear to be the amount of organic matter in the soil and the climatic conditions of the area (Lichtenstein et al., 1960).

The rates of loss of lindane were calculated by Wheatley (1965) in 10 long-term field studies in the United Kingdom. When lindane was applied to the soil surface, there was a 50% loss within 4–6 weeks and a 90% loss within 30–40 weeks. When the lindane was mixed into the soil, a 50% loss was observed after 15–20 weeks and 90% within 2–3 years. No lindane was recovered 13 years after the last application of lindane to a loam soil in Nova Scotia at a rate of 0.84–1.7 kg/ha (Stewart & Fox, 1971). Cliath & Spencer (1971) treated two test plots in California, USA, with 21 kg/ha, which is an application rate about 20 times above normal. A half-time of 8 months was found in sandy loam and 10 months in silty clay.

After application of lindane on three test plots of light sandy soil in the Netherlands for 15 years, to give total amounts of 6.5, 13.0, and 24.3 kg/ha, only 3, 4, and 8% of the applied amount, respectively, was recovered in layers to a depth of 20 cm (Voerman & Besemer, 1970). A further follow-up of these plots for four years showed rapid disappearance on the two locations with the lower application rates; slower degradation was seen on the plot that had received the highest application, where lindane was found to a depth of 40 cm (Voerman & Besemer, 1975). Admixture of a 5% lindane dust to the top 15-cm layer of a test plot at a rate of 10 kg of active ingredient per hectare in India led to an initial concentration of 3.2 mg/kg soil. After an observation period of 180 days, 97.7% of the applied lindane had disappeared. The initial half-time of lindane in this study was about 30 days (Agnihotri et al., 1977).

The degradation of gamma-HCH was also determined in a variety of studies in which technical-grade HCH was applied to soils. In most of these investigations, the application rates were extremely high, and in some, applications were made once a year for several years (Lichtenstein & Polivka, 1959; Stewart & Chisholm, 1971; Shiota & Kanda, 1972; Nash et al., 1973; Jackson et al., 1974; Suzuki et al., 1975). Under these conditions, gamma-HCH disappeared slowly from the soils and sometimes persisted for long periods.

The distribution of HCHs was studied in soil treated with BHC-20 (containing 70% alpha-HCH, 6.5% beta-HCH, 13.5% gamma-HCH, and 5% delta-HCH) in an agricultural area. The concentrations changed with time after application; the mean value for gamma-HCH was 16 µg/kg. The organic carbon content of the soil appeared to be of primary importance, and the significant decrease in isomer concentration observed with greater soil moisture was attributed to microbial degradation, which is favoured by these conditions (Chessells et al., 1988).

Kathpal et al. (1988) studied the behaviour of a formulation consisting of a mixture of five HCH isomers in paddy soils under subtropical conditions in India. The recommended application rate of 2.5 kg active ingredient per hectare and a rate of 5.0 kg/ha were used. Gamma-HCH had dissipated by 50–63% within three months under paddy, and average residues in soil at harvest were 0.3–0.34 mg/kg. Dissipation after nine months (two crop seasons) was 98%. The persistence under paddy in this study was fairly high, probably owing to the anerobic conditions, which slow microbial degradation. The paddy plants absorbed gamma-HCH from the soil: the residues at harvest were about 1.0 mg/kg in plants and 0.03–0.06 mg/kg in seeds.

4.2.3 Hydrolytic degradation

Determination of the hydrolytic stability of a substance provides an indication of whether this process can contribute to the disappearance of the substance from the aquatic environment and, to a certain extent, from soil. In a model experiment, the half-time of lindane at 22 °C was 47.9 h at pH 9 and 100.7 h at pH 7; no measurable hydrolysis occurred at pH 5 (Heupt, 1983).

4.2.4 Photolytic degradation (laboratory studies)

As lindane has measurable volatility and can be found at low levels in air, its degradation in sunlight has been studied.

Carbon dioxide was formed after ^{14}C-lindane was adsorbed onto silica-gel plates at a concentration of 33 µg/kg and irradiated with artificial sunlight (> 290 nm) in the presence of air; 6.4% of the carbon was oxidized within 17 h. This photo-induced oxidation was enhanced when the lindane was exposed to pure oxygen during irradiation (Kotzias et al., 1981). No measurable degradation (less than 0.5%) was observed 2000 h after exposure of lindane to the light of a Xenon lamp in a Xenotest 150 on the wall of a quartz vessel (solid phase). When the irradiation was performed in aqueous solution, about 4% of the applied lindane was degraded after 2000 h. The main degradation product was PCCH (Gardais & Scherrer, 1979).

Irradiation of lindane with ultra-violet light (254 nm) is obviously more effective for degradation of the compound than irradiation with light of longer wavelengths. Hamada et al. (1981, 1982) found rapid degradation of lindane in both the crystalline state and in solution with 2-propanol under these conditions, with PCCHs and TCCHs as reaction products. Eichler (1977) also found rapid degradation of lindane in the solid or gaseous form and in aqueous solution in the presence of ultra-violet irradiation, with half-times of 12–24 h for the first two phases and 1–2 days for the latter two.

4.2.5 Biodegradation in water

In a study of the degradation of lindane in a biological purification plant, 75% of the compound was degraded within 6 h (Eichler et al., 1976).

Newland et al. (1969) investigated the degradation of gamma-HCH in simulated lake impoundments. Sediments from Lake Tomahawk,

Wisconsin, USA, were added to solutions of 5 mg/litre ^{14}C-labelled lindane and equilibrated for 18 h, and aerobic and anaerobic tests were run for approximately 88 days. Initially, about 45% of the applied lindane was adsorbed to the sediment (200 g per 3-litre solution). Under aerobic conditions, about 16% of the added lindane was degraded by the end of the observation period, whereas more than 97% was degraded under anaerobic conditions. When lindane degradation was tested in samples of surface water from two different regions for periods of 3, 6, or 12 weeks, decreases of up to 90% of the initial concentration were found. Most of the lindane was metabolized by microorganisms in the sediments: In samples of sediment and water autoclaved prior to treatment and incubation, up to 95% of the applied lindane was still present (Oloffs et al., 1973).

In a field test in rice fields in the Camargue, France, a formulation containing lindane was applied at a rate that resulted in an initial concentration in water of 54.8 mg/m^3. Rapid disappearance was observed, for a half-time of about 1.5 days, and within 10 days the concentration had dropped to the background value of 0.08 mg/m^3 (Podlejski & Dervieux, 1978).

The degradation of lindane was also tested in the water of a drainage canal in the Holland Marsh, Ontario, Canada, in distilled water, and in both water types after sterilization. The half-time of lindane in the natural water was about six weeks, but a very low disappearance rate was seen in the distilled and sterilized water, indicating the importance of microbial action for degradation of lindane in water (Sharom et al., 1980).

An aquatic model ecosystem, with pond water, sludge, aquatic plants, and fish, was used to study the decomposition and migration of lindane. In water without hydrobionts, the half-time was 50 days. When sludge and aquatic plants were present, the half-time was 34 days, and that in the presence of fish was 2 days (Vrochinsky, 1973; see Izmerov, 1983).

4.2.6 Microbial degradation

A variety of experiments on the degradation of lindane was performed with mixed populations of the microorganisms that occur in different types of soil, in aquatic sediments (Newland et al., 1969; Benezet & Matsumura, 1973), and in other types of soil under aerated, submerged, and strictly anaerobic conditions (Macrae et al., 1967; Yule et al., 1967; Kohnen et al., 1975; Mathur & Saha, 1975, 1977; Tu, 1975; Haider, 1979). The fact that lindane was removed faster from non-sterile than from autoclaved soil demonstrated that its degradation in soil is due to microbial activity (Macrae et al., 1967; Kohnen et al., 1975).

The microorganisms shown by screening experiments to be capable of metabolizing and degrading lindane are as follows (Tu, 1976; Jagnow et al., 1977):

Bacteria	Fungi	Algae
Arthrobacter sp.	*Penicillium* sp.	*Chlamydomonas* sp.
Bacillus sp.	*Rhizopus* sp.	*Chlorella* sp.
Citrobacter sp.		
Clostridium sp.		
Enterobacter sp.		
Micromonospora sp.		
Pseudomonas sp.		
Thermoactinomycetes sp.		

In addition, lindane was metabolized in cell-free preparations of *Clostridium* sp. *in vitro* (Heritage & Macrae, 1977a; Ohisa et al., 1980).

Lindane is degraded by soil microorganisms under aerobic as well as under anaerobic conditions, but anaerobic conditions are the most favourable for its metabolism (Newland et al., 1969; Haider & Jagnow, 1975; Vonk & Quirijns, 1979). In an anaerobically grown culture of *Clostridium sphenoides* supplemented with lindane at 5 mg/litre, none was found, even after 2 h (Heritage & Macrae, 1979). Several species of soil bacteria that have been shown to degrade lindane effectively are described in detail in section 6.6.2.

In field studies in which gamma-HCH was applied at excessive doses, it was degraded more slowly than at doses closer to those used for normal agricultural applications. Introduction of HCH at up to 224 kg/ha, corresponding to 33.6 kg gamma-HCH per hectare, exceeded the degradation capacity of soil microorganisms for a long period (Nash et al., 1973). In addition, the analytical methods used might have resulted in an overestimation of the actual gamma-HCH concentration, as concluded by Lichtenstein & Schulz (1958b). Therefore, studies in which technical-grade HCH is applied, especially at excessive rates, cannot be used to evaluate the degradability of lindane in soil.

Environmental transport, distribution, and transformation

4.2.7 Bioaccumulation/Biomagnification

4.2.7.1 n-Octanol/water partition coefficient

The *n*-octanol/water partition coefficient (P_{ow}) of lindane was determined in several studies, with good agreement, covering the narrow range of log P_{ow} = 3.29–3.72 (Kurihara et al., 1973; Platford, 1981; Darskus, 1982; Geyer et al., 1982; Hermens & Leeuwangh, 1982; Geyer et al., 1984). These values indicate that lindane can become enriched in lipid-containing biological compartments.

4.2.7.2 Aquatic environment

The bioconcentration factor for lindane was found to be dependent on the concentration to which the organisms, such as algae, crustaceae, and fish, were exposed (Bauer, 1972; Ernst, 1975; Schimmel et al., 1977; Trautmann & Streit, 1979; Marcelle & Thome, 1983): The highest bioconcentration factors were seen with the lowest exposure concentrations. For example, Marcelle & Thome (1983) determined the residues of lindane in brain, liver, and muscle of the gudgeon (*Gobio gobio*) after exposure to concentrations of 0.22–142 µg/litre lindane in water. At the lowest concentration, the bioconcentration factors in brain, liver, and muscle were about 600, 200, and 100, respectively, but they decreased to values of less than 50 at higher concentrations.

Mouvet (1985) transplanted the freshwater aquatic moss *Cinclidotus danubicus* from an uncontaminated area to a river that received the effluent from an insecticide factory and determined gamma-HCH concentrations in water and moss 13, 24, and 51 days after the transplant. A three-fold increase in the gamma-HCH level was found, with a bioconcentration factor of 294.

In a variety of aquatic organisms exposed to contaminated water, the bioconcentration factor for lindane ranged from 13 to 1000 on a wet weight basis (Table 3).

Another approach to the study of the bioconcentration of lindane is the use of systems that simulate natural conditions, taking into account sedimentary absorption processes and the influence of contaminated food. The bioconcentration factors for brine shrimp, mosquito larvae, and the brook silverside (*Haludesthes sicculus sicculus*) exposed to lindane applied

Table 3. Bioconcentration factors of lindane in laboratory experiments; test organisms were exposed to contaminated water for the specified time

System	Exposure time	Exposure concentration (µg/litre)	Bioconcentration factor[a]	Reference
Algae				
Cladophora sp.	up to 48 h	80.0 3.9	180 (d) 341 (d)	Bauer (1972)
Nitzschia actinastroides	24 h	6.1	1500–4700 (v) 4400–12 400 (d)	Trautmann & Streit (1979)
Molluscs				
Aplysia punctata	3-6 days	9000	201–436 (w)	Chabert & Vicente (1978)
Mya arenaria Mercenaria mercenaria	5 days 5 days	5	40 13	Butler (1971)
Mytilus edulis	ns	2.61 0.02	74 (w) 242 (w)	Ernst (1975)
Mytilus edulis	ns	2–5	139 (w)	Ernst (1979)
Venerupis japonica	3 days	1	121 (ns)	Yamato et al. (1983)
Annelidae				
Lanice conchilega	ns	2–5	1240 (w)	Ernst (1979)
Crustacea				
Penaeus duorarum Palaemonetes pugio	96 h 96 h	0.23 1.0	143 (ns) 80 (ns)	Schimmel et al. (1977)
Insects				
Sigara striata and Sigara lateralis	1 day	10	70–100	Kopf & Schwoerbel (1980)
Fish				
Lagodon rhomboides	96 h	23.0	287 (ns)	Schimmel et al. (1977)
Cyprinodon variegatus	96 h	108.7	727 (ns)	
Leuciscus idus Cyprinus carpio Salmo truttafario Poecilia reticulata	ns	10–500	765 (ns) 281 (ns) 442 (ns) 938 (ns)	Sugiura et al. (1979)
Poecilia reticulata	4 days	1	697 (ns)	Yamato et al. (1983)
Salmo gairdneri	27 days	30–290	319	Ramamoorthy (1985)

[a] Calculated on the basis of: wet weight (w), dry weight (d), volume (v); ns, not specified

to the sand of a test aquarium were 95, 220–383, and 600–1613, respectively, depending on the food chains used (Matsumura & Benezet, 1973). Marcelle & Thome (1984) investigated the bioconcentration of lindane in the gudgeon (*Gobio gobio*) in relation to the route of exposure. Fish were exposed either to contaminated water alone or additionally to contaminated food. After 18 days, the group fed contaminated food had a 2.5-fold higher level of lindane residues in liver than the group exposed to contaminated water alone. Within three days after cessation of exposure, 98.4% of the lindane residues had been excreted.

The uptake, transport, and bioconcentration of lindane were also studied in a freshwater food chain, which consisted of *Chlorella* sp., *Daphnia magna*, and *Gasterosteus aculeatus* (algae–crustacea–fish). Uptake from water was more rapid than uptake from food and depended on the duration of the experiment and the feeding rate. The increase in lindane residues in the last link of the food chain (fish) was not directly proportional to the concentration found in the primary links (Hansen, 1980).

4.2.7.3 Terrestrial environment

The bioconcentration of lindane was investigated in a terrestrial food chain, which consisted of soil, barley, caterpillar, and quail. Doses up to 400 times the standard agricultural dose (50 and 200 mg/kg soil) were applied to the soil. Although lindane was found in all of the links of the food chain, the concentrations decreased progressively (Dugast, 1980).

Feeding hens diets containing lindane at 0.05, 0.15, or 0.45 mg/kg for 20 weeks resulted in constant values of 0.01, 0.03, and 0.09 mg/kg of eggs, demonstrating a dose-related accumulation of lindane (Cummings et al., 1966).

Several studies are available on the bioconcentration of lindane in rats. After seven rats had received daily doses of 2 or 4 mg/kg body weight for up to 12 weeks, gamma-HCH was found at a concentration of about 8 mg/kg in adipose tissues of the group that had recived the high dose (Jacobs et al., 1974). In another experiment, four generations of rats were fed a diet containing 20% fat and a mixture of insecticides including lindane at levels of 0.07–0.8 mg/kg. Even in the F_3 generation, the levels of gamma-HCH residues in adipose tissues were of the same order of magnitude (< 0.05–0.56 mg/kg) as those of lindane in the diet (Adams et al., 1974). No accumulation occurred, even in four consecutive generations.

Accumulation factors have been determined from feeding studies in rats (Fitzhugh et al., 1950; Oshiba, 1972; Baron et al., 1975; Suter et al., 1983). In comparison to the concentration of lindane in the diet, the highest reported bioconcentration factor was about 2 for adipose tissue. The average bioconcentration factor for adipose tissues in rats derived from all these studies is 1; the bioconcentration factors for other tissues are considerably lower.

4.2.7.4 Bioconcentration in humans

Geyer et al. (1986) examined data on environmental chemicals detectable in adipose tissue and/or breast milk of non-occupationally exposed humans and concluded that, in industrialized countries, more than 90% of human exposure to HCHs originates from food. Mean concentrations of gamma-HCH in human adipose tissue in Czechoslovakia, the Federal Republic of Germany, and the Netherlands were 0.086, 0.024–0.061, and 0.01–0.02 mg/kg, respectively, on a fat basis. The mean bioconcentration factor, calculated on the basis of the concentration in the diet (2.3, 5.0, and 0.62–0.9 µg/kg, respectively) and levels in adipose tissue, was 18.6 ± 9 on a lipid basis (range, 10.4–32.5). Greve & Wegman (1985) found an accumulation factor (adipose tissue/blood) of 70 for gamma-HCH in humans.

4.2.7.5 Field studies

The bioconcentration of lindane was investigated by environmental monitoring in aquatic ecosystems. The residue levels found in different organisms were related to the environmental background levels, and these data were used to calculate the bioconcentration factors.

The bioconcentration factor for gamma-HCH in sea water and bladder wrack (*Fucus vesiculosus*) in the Husum estuary and the adjacent North Friesian Wadden Sea in the Netherlands was about 150 (Herrmann et al., 1984). On the basis of the data given in section 5.1.5.2 on the occurrence of gamma-HCH in muscle and fat of bream collected in the River Elbe, a bioconcentration factor of 10 000 to 50 000 was calculated (Arbeitsgemeinschaft für die Reinhaltung der Elbe, 1982).

Frisque et al. (1983) studied the accumulation of lindane by bryophytes (*Cinclidotus danubicus* and *C. nigricans*) in the Meuse River and found a concentration factor of 300–350. The average level in the river was

0.067 µg/litre. Hartley & Johnston (1983) found a bioconcentration factor for the freshwater clam *Corbicula manilensis* of 2610 on a lipid basis; and Cosson Mannevy & Marchand (1980) found a mean factor of 26 198 (on a dry-weight basis) in *Mytilus edulis*.

On the basis of the concentrations of gamma-HCH in sea water, sediments, and fish from the Mediterranean Sea, El-Dib & Badawy (1985) calculated a bioconcentration factor of about 1000 (on a lipid basis). Tanabe et al. (1984) reported bioconcentration factors for total HCHs in a trophic chain in the western North Pacific. As the contribution of gamma-HCH to the residue levels was determined, the bioconcentration factors for this isomer can be estimated to be about 40, 40, 100, and 1850 for zooplankton, myctophid, squid, and dolphin, respectively.

5. ENVIRONMENTAL LEVELS AND HUMAN EXPOSURE

5.1 Environmental levels

5.1.1 Air

An average of 0.23 ng/m³ (0.039–0.68 ng/m³) gamma-HCH was found in 24 samples of air taken from over the western Pacific, the eastern Indian Ocean, and the Antarctic Ocean (Tanabe et al., 1982).

Levels of gamma-HCH in the air of various regions of the USA were within a similar range (US Environmental Protection Agency, 1976). The levels were below 1 ng/m³ in most samples, and values up to 16.2 ng/m³ were found in only two regions.

Gamma-HCH was found at an average concentration of 0.14 ng/m³ in the air of unpolluted areas in the Federal Republic of Germany in 1972; in polluted areas (the Ruhr), a level of 0.8 ng/m³ was found in 1976/77 (Deutsche Forschungsgemeinschaft, 1983; Hildebrandt et al., 1986). It occurred at 0.52–11 ng/m³ in a location with heavy traffic near Ulm in the Federal Republic of Germany and at 0.18–1.1 ng/m³ in a rural area. The authors concluded that the concentrations in the lower troposphere under different meteorological conditions reflect regional input and long-range transport (Wittlinger & Ballschmiter, 1987).

The average concentration of gamma-HCH in 55 air samples taken near Delft, the Netherlands, in 1979–80 was 0.36 ng/m³ (maximum, 3.4 ng/m³); in three other locations in the Netherlands, average levels of 0.2–0.9 ng/m³ were found. In six houses built on former dumping grounds, the average concentration of gamma-HCH was 6 ng/m³ (range, 1–14 ng/m³), whereas in the space beneath the floor the level was below the detection limit (1 ng/m³). Outdoor concentrations in this area were 0.3–0.4 ng/m³. In another study, the concentrations of gamma-HCH in the space beneath the floor of houses were 90 ng/m³. Much higher levels were found in houses treated with lindane-containing products for the control of woodworm or of long-horned beetle. Peak levels of 51–61 µg/m³ were found four weeks after application; these decreased gradually to 8–24 µg/m³ after 10 weeks. After indoor application of lindane for wood preservation, levels of 50 µg/m³ were common, with peak levels of up to 100 µg/m³ (Slooff & Matthijsen, 1988).

Environmental levels and human exposure

5.1.2 Water

5.1.2.1 Rain and snow

Levels of 0.001–0.005 µg/litre were found in rain-water analysed in the Federal Republic of Germany in 1970–72 (Mestres, 1974); in 1983, gamma-HCH was found at an average of 0.06 (range, 0.01–0.18) µg/litre in rain-water near de Bilt, the Netherlands (Slooff & Matthijsen, 1988).

Strachan et al. (1980) found traces of gamma-HCH in 17 samples of snow collected from the Canadian side of the Great Lakes in 1976 and 5–12 ng/litre in 81 samples of rain-water collected in 1976 and 1977.

5.1.2.2 Fresh water

Water samples from selected rivers in Yorkshire, United Kingdom, analysed for gamma-BHC in 1966 contained levels of 0.001–0.18 µg/litre; in 1968, however, the highest value was 0.622 µg/litre. Water samples from six other rivers, also analysed in 1968, contained mean values of 0.011–0.030 µg/litre, and the highest levels found were 0.020–0.098 µg/litre (Lowden et al., 1969).

River water samples analysed in 1969–72 in Belgium, France, the Federal Republic of Germany, the Netherlands, and Italy contained less than 0.1 µg/litre and usually less than 0.05 µg/litre. In 1826 water samples taken at 99 sites in the Netherlands in 1966–77, the highest concentrations of gamma-HCH were found in those from the River Rhine and its tributaries. The concentrations of gamma-HCH over the period 1969–74 varied between 0.01 and 0.4 µg/litre, but in 1974–77, the concentrations were all below 0.1 µg/litre (Mestres, 1974). Gamma-HCH concentrations have been measured in the Rivers Rhine, Meuse, and West-Scheldt and in other surface waters in the Netherlands since 1969. Since 1974–75, the levels have been below 0.05 µg/litre in the Rhine and about 0.05 µg/litre in the West-Scheldt; in the Meuse, the concentrations were more variable and ranged from 0.01 to 1.0 µg/litre. In agricultural and horticultural areas, the levels were 0.01–1.0 µg/litre, with incidental peaks up to 0.5 µg/litre, probably due to use of lindane. The average concentration of dissolved gamma-HCH in the Meuse–Rhine estuary in 1974 was 20 ng/litre and that of suspended gamma-HCH between 1 and 20 ng/litre. In coastal waters of the Netherlands, the concentration of dissolved gamma-HCH was 0.9–4.6 ng/litre and that of bound gamma-HCH, 3.1–8.7 ng/litre (Slooff & Matthijsen, 1988).

A sampling trip along the River Rhine, from Rheinfelden in Switzerland to Rotterdam in the Netherlands, proved that the source of alpha-, beta-, and gamma-HCH was located in the upper reaches of the River. In the Meuse, lindane levels in 1969–77 were all below 0.1 µg/litre (Wegman & Greve, 1980). In an extensive programme in 1982 to determine pollution in Dutch surface waters at 45 locations, gamma-HCH concentrations were generally between 0.01 and 0.1 µg/litre (Wammes et al., 1983).

The mean concentration of gamma-HCH in the River Elbe, from Schnackenburg to the North Sea, in 1981–82 was 0.021 (< 0.001–0.051) µg/litre; during February–November 1988, the concentrations were 0.005–0.044 µg/litre (Arbeitsgemeinschaft für die Reinhaltung der Elbe, 1988). More figures for Germany are given by Wirth (1985). Gamma-HCH was found at three locations in the River Rhine at 0.02 µg/litre and in six side-rivers at 0.01–0.06 µg/litre. These levels had decreased markedly since 1975 (Landesamt für Wasser und Abfall, 1988).

5.1.2.3 Sea water

Atlas & Giam (1981), Bidleman & Leonard (1982), Oehme & Stray (1982), and Oehme & Mano (1984) analysed water from such widely differing areas as the Eniwetok Atoll in the North Pacific, the Arabian Sea, the Persian Gulf, the Red Sea, Lillestrøm, Norway, Bear Island, and Spitzbergen in the Arctic Ocean. The gamma-HCH concentrations were in the range 0.01–0.05 ng/litre, except in the Arabian Sea, the Persian Gulf, and at Lillestrøm, where levels up to 0.67 ng/litre were found (Slooff & Matthijsen, 1988). Levels of 0.0001–0.004 µg/litre gamma-HCH were measured in the Western Pacific, the Eastern Indian, and Antarctic Oceans (Tanabe et al., 1982). No gamma-HCH was found in 60 water samples from the Japan Sea and Pacific Ocean (detection limit, 0.1 µg/litre) (A. Hamada, letter to M. Mercier, dated 28 July 1989; T. Onishi, letter to M. Mercier, dated 24 July 1989). The levels detected in water from the North Sea and the Arctic Sea are of the order of 0.001–0.02 µg/litre (Deutsche Forschungsgemeinschaft, 1983). The maximal level of gamma-HCH in North Sea water in 1972 was 0.028 µg/litre; 5–10% of the samples contained gamma-HCH (Mestres, 1974). The level of gamma-HCH in surface-water of the North Sea in June–July 1986 ranged from 1.0 to 4.0 ng/litre. The highest concentrations were found close to the coast (Umweltbundesamt, 1988–89).

Environmental levels and human exposure

5.1.3 Soil

Traces of gamma-HCH are transmitted to soil by precipitation; the resulting contamination is generally below the limit of detection (0.0001–0.001 mg/kg). Application of lindane in agricultural areas can result in higher concentrations: levels in some German districts were mainly in the range 0.001–0.01 mg/kg, but in certain fields up to 0.6 mg/kg was found (Fricke, 1972).

Edelman (1984) analysed 96 samples of the upper 10 cm of soil from 38 natural reserves in the Netherlands for gamma-HCH: 59 samples contained < 1 µg/kg, 21 contained 1–10, 9 had 10–20, and 7 had 20–80 µg/kg (Slooff & Matthijsen, 1988). In the National Soils Monitoring Program of the US Environmental Protection Agency (Carey et al., 1979), several thousand samples from cropland sites were analysed for residues; no gamma-HCH residues were detected in more than 99% of the samples. In the Ukraine, however, 36 of 136 soil samples taken at various locations contained lindane at levels of 0.1–5 mg/kg (Talanov, 1977; see Izmerov, 1983).

In a study on the application of lindane dust by aircraft on mosquito breeding sites at 1.3 kg/ha, the gamma-HCH content of the soil was 1 mg/kg; after one year, the level was 0.01 mg/kg (Vroschinsky, 1973; see Izmerov, 1983).

5.1.3.1 Sediment

Gamma-HCH was present in three of six samples of sediment taken from Nyumba Ya Mungu Lake in the United Republic of Tanzania in 1986, at a concentration of 1–4 µg/kg dry weight (Paasivirta et al., 1988).

Martin & Hartmann (1985) found gamma-HCH at levels above the detection limit (5 µg/litre) in less than 4% of 117 samples of sediment taken in 1980–82 from riverine and pothole wetlands in north-central USA. In less than 4% of the samples, gamma-HCH was present at above the detection level of 5 µg/kg.

In Japan, gamma-HCH was found in 9 out of 60 samples of sediment at a concentration of 10 µg/kg in 1974 (A. Hamada, letter to M. Mercier, dated 28 July 1989; T. Onoshi, letter to M. Mercier, dated 24 July 1989).

The median levels of gamma-HCH in sediments from eight rivers, harbours, and sites close to dumping places in the Netherlands were 15–342 µg/kg dry matter (Slooff & Matthijsen, 1988).

5.1.3.2 Dumping grounds and sewage sludge

The soil at various locations in the Netherlands is polluted with HCHs as a result of spillage during production, storage, and handling of this chemical during the 1950s. The concentrations found range up to a few thousand milligrams of HCHs per kilogram of dry soil. Further pollution has been caused by the dumping of chemical waste, sometimes in order to level the ground; this waste can be dispersed from dumping areas by leaching or wind erosion. In certain polluted areas, high concentrations of HCHs (mainly alpha- and beta-HCHs) were found at depths of more than 2 m below ground level. In 18 locations in the Netherlands, the average concentrations of gamma-HCH in sewage sludge in 1981 were 8–50 µg/kg dry matter. Groundwater was also found to be polluted, but this was restricted to the vicinity of the production areas; horizontal transportation of HCHs in groundwater appeared to be limited (Slooff & Matthijsen, 1988).

Fieggen (1983) found gamma-HCH in sewage sludge at mean values of 25 µg/kg dry matter in 1975, 43 µg/kg in 1978, and 12 µg/kg in 1981.

5.1.4 Drinking-water, food and feed

Although in most countries nowadays only lindane is used, residues of alpha- and beta-HCH can still be found in crops and animal products originating from regions where technical-grade HCH (containing all of the HCH isomers) is still in use.

5.1.4.1 Drinking-water

Gamma-HCH was found at 0.0001–0.001 µg/litre in water from 19 lakes in Germany and at levels below 0.001 µg/litre (0.0001–0.0008 µg/litre) in the drinking-water derived from them (Bernhardt & Ziemons, 1974). In the USA, only 3% of drinking-water samples examined contained gamma-HCH, in a range of 0.001 to about 0.1 µg/litre (US Environmental Protection Agency, 1976). In Ottawa, Canada, drinking-water samples collected in 1976 contained 0.4–11 ng/litre (Williams et al., 1978).

5.1.4.2 Cereals, fruits, pulses, vegetables, and vegetable oil

The large body of information on gamma-HCH residue levels in crops grown and treated with this chemical according to Good Agricultural

Environmental levels and human exposure

Practice has been reviewed comprehensively by the FAO/WHO Joint Meeting on Pesticide Residues and summarized in published monographs (FAO/WHO, 1967, 1968, 1969, 1970, 1974, 1975, 1976, 1978, 1980).

In samples of ready-to-eat foods collected from 30 markets in 27 US cities in 1966–67, gamma-HCH levels were 0.003–0.009 (occasionally 0.06) mg/kg in grains and cereals, 0.002–0.027 mg/kg in garden fruits, 0.001–0.005 mg/kg in potatoes, 0.002–0.007 mg/kg in leafy vegetables, and 0.004–0.012 mg/kg product in oils, fat, and shortening (Martin & Duggan, 1968). In 1967–68, residues of gamma-HCH were found at 0.002–0.006 in leafy and root vegetables, at 0.002–0.003 in garden fruits, and at 0.029–0.085 mg/kg product in oils, fat, and shortening (Corneliussen, 1969).

In monitoring studies carried out on grain in the Federal Republic of Germany at one-year intervals since 1975, gamma-HCH residues in wheat and barley were 0.001 mg/kg or less (Ocker, 1983). More than 800 samples of cereal and cereal products analysed in Germany in 1975–78 and 1979–83 contained mean concentrations of 0.0009–0.04, but cereal products had up to 0.11 mg/kg. The mean concentration of gamma-HCH in 200 samples of wheat and rye collected in 1986 and 1987 was 0.06 mg/kg, with a maximum of 0.3 mg/kg (Umweltbundesamt, 1988–89).

Of 281 samples of wheat analysed for the presence of gamma-HCH in the United Kingdom between October 1978 and April 1979, 71 contained levels in the range 0.002–0.04 mg/kg. Gamma-HCH was also found in one sample of polished rice from Spain, at a concentration of 0.008 mg/kg (Steering Group on Food Surveillance, 1982). Gamma-HCH was found in 16% of samples of imported maize in the United Kingdom in the range none detected to 0.007 mg/kg, and in 28 samples of different types of pulses at none detected to 0.05 mg/kg. Of retail cereal products, only bran and wheat contained detectable levels (0.01 mg/kg product) of gamma-HCH in 1982 (Steering Group on Food Surveillance, 1986). In 1986–87, 31 of 142 samples of pulses contained residues; in nine, levels of < 0.01–0.4 mg/kg were found. Peanut butter and vegetable oils contained 0.01 mg/kg (Steering Group on Food Surveillance, 1989).

About 80–90% of samples of fruit, potatoes, and other vegetables analysed in the Federal Republic of Germany contained no detectable residues of gamma-HCH (Weigert et al., 1983). The remaining 10–20% had mean levels up to 0.01 mg/kg, with no significant difference between 360 samples originating from conventional agriculture and 360 samples from 'alternative' agriculture (Vetter et al., 1983). In 1976–78 and 1980, the mean concentrations of gamma-HCH were < 0.001–0.002 mg/kg product in more than 400 samples of fruit, potatoes, and other vegetables. In the Netherlands, residues in fruit and vegetables were generally in the range 0–0.1 mg/kg, although some leafy crops, such as endive, lettuce, celery, and leek, contained levels up to 5 mg/kg. Samples of wheat contained only 0–0.05 mg/kg, with a few measurements up to 0.2 mg/kg (FAO/WHO, 1978). In France, gamma-HCH residues were found in wheat at 0.01–0.02 mg/kg, and at low levels in other commodities, such as carrots and endives (Laugel, 1981). Engst et al. (1967) found that the gamma-HCH content of carrots grown from seed treated with this compound decreased continuously during the first 120 days. At normal harvesting time, the early varieties contained 3–6 mg/kg product, the mid-season varieties about 2 mg, and the late varieties, 0.4 mg/kg. When the carrots were harvested after 200 days, 0.3–0.7 mg/kg was present (independently of variety). Even after 6 months' storage, low residues were still present.

5.1.4.3 Meat, fat, milk, and eggs

Martin & Duggan (1968) found gamma-HCH at levels of 0.09 mg/kg in dairy products and at 0.01–0.03 mg/kg (with a peak of 0.374 mg/kg) in samples of meat, fish, and poultry collected from 30 markets in 27 cities in the USA in 1966–67. Residue levels in samples of meat, fish, and poultry in 1967–68 were 0.003–0.026 mg/kg (Corneliussen, 1969). No gamma-HCH or levels of 0.01–0.1 mg/kg were found in 99% of samples of cow's milk and manufactured milk products from Illinois (USA) (Wedberg et al., 1978). In milk samples collected during Spring 1983 from 359 bulk transporters, representing 16 municipalities of Ontario, Canada, gamma-HCH was found in 68% of the samples at a mean concentration of 4.0 µg/kg butter fat (Frank et al., 1985). Six samples of cow's milk from six locations in Switzerland contained 3.0–5.1 mg/kg on a fat basis (Rappe et al., 1987).

In about 25% of 976 samples of meat and poultry products (including eggs) collected in the United Kingdom in 1984–86, gamma-HCH was present at a mean concentration of 0.01–0.02 mg/kg. The highest level, 3.7 mg/kg, was found in lamb. Processed meat and poultry products (631 samples collected in 1985–87) contained mean concentrations of 0.01–0.06 mg/kg product. About half of 849 samples of retail milk and dairy products collected in 1984–87 contained gamma-HCH at concentrations of 0.01–0.03 mg/kg; the highest level, 0.7 mg/kg, was found in milk (Steering Group on Food Surveillance, 1989). Imported meat products were also analysed in the United Kingdom for the presence of alpha-, beta-, and gamma-HCH. No detectable residue of gamma-HCH was found in beef or pork products: processed pork contained none detectable to 0.03 mg/kg.

Environmental levels and human exposure

Processed poultry contained none detectable to 0.04 mg/kg (Steering Group on Food Surveillance, 1986). In 1967–70, in the Ukraine, gamma-HCH was found in cows' milk at an average concentration of 0.6 mg/litre (Medvedev & Perepechkina, 1973; see Izmerov, 1983). In the USSR, the following concentrations were found: milk and milk products, 0.055 ± 0.005 mg/kg; poultry and fish, 0.068 ± 0.021 mg/kg; butter, 0.003 ± 0.002 mg/kg; vegetables and fruits, 0.008 ± 0.003 mg/kg; groats and flour, 0.005 ± 0.002 mg/kg (Sizova & Bogomolova, 1976; see Izmerov, 1983).

Concentrations of gamma-HCH were measured in 1250 samples of milk and other dairy products in France in 1970–77 and in 1981. In the first period, the gamma-HCH concentration was < 0.1 mg/kg of fat; by 1981, the levels had declined to < 0.03 mg/kg of fat (Laugel, 1981; Rhône-Poulenc Agrochimie, 1986). Higher levels (mean, 0.85 mg/kg) were found in animal fat, but meat and eggs generally contained no detectable residue (Laugel, 1981). The mean levels of gamma-HCH found in a large number of samples of various food items in Germany (Hildebrandt et al., 1986) are shown in Table 4.

The levels of gamma-HCH in food items analysed in France were 0.006–0.01 mg/kg in 113 samples of vegetables, 0.005–0.04 mg/kg in 192 samples of fish and seafood, 0.005–0.041 mg/kg in 154 samples of preserved meat, 0.007–0.017 mg/kg in 104 samples of cereal products, 0.007–0.034 mg/kg in 120 samples of butter and cheese, 0.005–0.059 mg/kg in 25 samples of oil and fat, and 0.006–0.021 mg/kg in 26 samples of fruit (Rhône-Poulenc Agrochimie, 1986).

Skaftason & Johannesson (1979) found a mean value of 13 µg/kg in 35 samples of butter from Iceland in 1968–70. Of 32 samples analysed in 1974–78, only five contained gamma-HCH, at a mean value of 7 ± 2 µg/kg. The mean concentration in meat, poultry and eggs in the Netherlands in 1976–78 was 0.002 mg/kg (range, 0.001–0.004 mg/kg) (De Vos et al., 1984); the levels in dairy products were similar.

Fifteen of 105 chicken eggs from seven areas in Kenya had a median concentration of 0.01 mg/kg (range, 0.01–0.04 mg/kg) (Kahunyo et al., 1988). Ten samples each from two lots of lamb and beef were collected randomly from markets in Bagdad, Iraq, in 1983 and analysed for the presence of gamma-HCH. An average concentration of 0.225 (0.004–1.611) mg/kg was found in lamb, and 0.116 (0.005–0.83) mg/kg was found in beef (Al-Omar et al., 1985).

Table 4. Levels of gamma-HCH (mg/kg) in food items in Germany

Food item	1973–78	1979–83	1973–83
Meat[a]			0.004–0.04
Meat products[a]		0.006–0.055 (maximum, 0.52)	
Animal fat[a]			0.007–0.09 (maximum, 0.5)
Game[a]			0.042–4.072
Poultry[a]	0.01–0.05	0.004–0.046 (maximum, 0.471)	
Chicken eggs Chicken eggs[a,c]	0.001–0.02 (maximum, 1.9)		< 0.001–0.01
Milk and milk products[a]	0.05	0.01–0.02	
Cow's milk[a,b]	0.03	0.01	
Vegetable oil and margarine[a]	0.01–0.02		
Oil seeds, nuts, pulses		0.001–0.127	
Fish and fish products	0.01–0.02	0.002–0.009	
Shell-fish and molluscs		< 0.001–0.020	

[a] From Hildebrandt et al. (1986); on fat basis
[b] From Anon. (1984)
[c] From Koelling (1978)

5.1.4.4 Animal feed

Of 114 samples of animal feed analysed in the United Kingdom in 1982–85, 49 contained gamma-HCH at concentrations up to 2.3 mg/kg product (Steering Group on Food Surveillance, 1986).

5.1.4.5 Miscellaneous products

Lanolin produced from crude wool grease may contain gamma-HCH: a level of 1.2 mg/kg was found in the USA (Anon. 1989); and Meemken et al. (1982) found average levels of 2.4 and 2.1 mg/kg in 1976 and 1981,

Environmental levels and human exposure

respectively, in Germany. Concentrations of 0.001–0.23 mg/kg were found in cosmetic creams made from the lanolin.

5.1.5 Terrestrial and aquatic organisms

5.1.5.1 Plants

Gamma-HCH was present in most of 13 samples of three types of moss and four types of lichen collected on the Antarctic Peninsula (Graham Land) in 1985 at a mean concentration of 0.84 mg/kg (range, 0.4–1.7 µg/kg) (Bacci et al., 1986).

In 1984, near Florence and Siena, Italy, far from primary sources of pollution, leaves from ten species of tree and two species of lichen were found to contain average levels of 8.2 (range, 2–14) and 10 (9–11) µg/kg dry weight, respectively. Gamma-HCH levels in plant species collected in 14 countries ranged from 0.2 to 700 µg/kg dry weight (Gaggi et al., 1986).

5.1.5.2 Aquatic organisms

Freshwater mussels (*Anodonta piscinalis*) were used to monitor bioaccumulation of pollutants at 17 sampling sites in a river basin in Finland between 1984 and 1987. One to three mussels were used per sampling site. Gamma-HCH was found in concentrations varying from none detected to 553 µg/kg on a fat basis; however, a clear decrease was seen over the period of study (Herve et al., 1988).

Cowan (1981) studied the extent of pollution by HCHs of Scottish coastal waters using *Mytilus edulis* as the biological indicator. The gamma-HCH levels at 118 sites were < 6–53 µg/kg dry weight, which are similar to those found in Germany, the Netherlands, Spain, and the United Kingdom. The fish and shellfish sampling programme of the Ministry of Agriculture, Fisheries, and Food in the United Kingdom in 1977–84 was implemented mainly in areas around the coasts of England and Wales. The range found for gamma-HCH was < 0.001 (none detected) to 0.075 mg/kg wet weight; the level in fish muscle was < 0.001 mg/kg wet weight (Franklin, 1987).

The average concentration of gamma-HCH was measured in 10 marine organisms collected along the Mediterranean coast of Spain during 1985. *Mytilus galloprovincialis*, *Venus gallina*, *Sardina pilchardus*, and *Mullus surmuletus* contained 0.1–1.7 µg/kg fresh weight (maximum, 16 µg/kg) (Pastor et al., 1988).

Bream collected in rivers and lakes at 15 locations in Germany contained average gamma-HCH concentrations of 106–696 µg/kg on a fat basis (Umweltbundesamt, 1988–89), while bream collected in the River Elbe, between Schnackenburg and the North Sea, contained average concentrations up to 0.031 mg/kg in muscle and up to 2.6 mg/kg in adipose tissue (Arbeitgemeinschaft für die Reinhaltung der Elbe, 1982). In 1970–72, different types of fish, mussels, and shrimps were analysed for gamma-HCH. Fish with a low fat content, collected in the Atlantic Ocean and the North Sea, contained 0.004–0.008 mg/kg fresh weight, and fat fish contained 0.01 (0.01–0.026) mg/kg fresh weight. Fat fish caught in the Baltic contained higher levels—up to 0.2 mg/kg fresh weight. Mussels and shrimp caught in the North Sea contained none to 0.009 mg/kg fresh weight; mussels from the Baltic coast contained 0.009–0.011 mg/kg. In 1973–76, similar values were found, except that the fat fish had lower levels. Marine organisms from the Baltic Sea generally contained higher levels of gamma-HCH than those from the North Sea. Freshwater fish from industrially contaminated areas contained higher levels (Hildebrandt et al., 1986). Gamma-HCH was detected at levels up to 7.0 µg/kg (mean, 2.5 µg/kg) in the muscle of flounders collected off the coast of the North Sea in Germany in 1986 (Umweltbundesamt, 1988–89).

5.1.5.3 Terrestrial organisms

Earthworms: Gamma-HCH was found in the soil of ten arable and two orchard sites in the United Kingdom at 0.01 and 0.08 mg/kg soil, respectively, and in worms living in the two soils at 0.05 and 0.3 mg/kg (Advisory Committee on Pesticides and Other Toxic Chemicals, 1969).

Birds: Bednarek et al. (1975) determined total HCH isomers at levels of 0.03–0.63 mg/kg total egg (or 0.6–11.1 mg/kg on a fat basis) in eggs of birds of prey, such as the sparrowhawk (*Accipiter nisus*), in two areas of Germany in 1972–73. Eggs of sandwich terns collected in the Elbe estuary contained arithmetic means (10 eggs) of 0.006 mg/kg fresh weight in 1981, 0.002 in 1985, 0.003 in 1986, and 0.028 in 1987 (Umweltbundesamt, 1988–89). The concentrations found in the livers of avian predators in the United Kingdom are shown in Table 5.

The mean levels of gamma-HCH detected in 23 barn owls (*Tyto alba* Scop.) obtained in Leon, Spain, were 0.03 (0.003–0.083) mg/kg wet weight in muscle, 0.036 (0.002–0.208) in liver, 0.051 (0.009–0.144) in fat, 0.012 (0.002–0.031) in brain, and 0.081 (0.005–0.343) in kidneys (Sierra & Santiago, 1987).

Table 5. Residues of gamma-HCH in livers of avian predators in the United Kingdom[a]

Bird	Date	No. of samples	gamma-HCH (mg/kg)
Sparrowhawk	1963	11	0.01
Kestrel	1963	20	0.04
	1964	28	0.1
	1965	60	0.03
Tawny owl	1964	14	0.01
Heron (adults)	1964	17	0.005
Great crested grebe	1963/66	15	0.03

[a]From Advisory Committee on Pesticides and Other Toxic Chemicals (1969)

Faladysz & Szefer (1982) examined adipose tissue from seven species of diving ducks at their winter quarters in the southern Baltic. Residues of gamma-HCH were detected in only 4 of 129 samples from three species of duck examined (range, 0.001–0.51 mg/kg on a fat basis).

Mammals: No gamma-HCH (< 0.01 mg/kg) was found in muscle tissue from 51 North American wolves captured in 1969–71 in sparsely populated forest regions (Schneeweis et al., 1974). Norstrom et al. (1988) determined the contamination of the marine ecosystem of the Canadian Arctic and sub-Arctic by organochlorine compounds by analysing adipose tissue and liver from 6–20 polar bears (*Ursus maritimus*) per zone collected from 12 zones between 1982 and 1984. The levels were 0.30–0.87 mg/kg on a fat basis; the highest levels were found in zones receiving continental run-off.

Mean concentrations of gamma-HCH in 86 samples of kidney fat from roe-deer (*Capreolus capreolus*) collected in five locations in Germany in 1985–86 were 8–12 µg/kg, with a maximum of 1020 µg/kg (Umweltbundesamt, 1988–89).

5.2 Exposure of the general population

The data presented above demonstrate that the main source of exposure of the general population is food.

5.2.1 Total-diet studies

In total-diet studies carried out in the United Kingdom between 1966 and 1985, 22–25 samples of foods in 20–24 groups were purchased in 21 towns throughout the country and prepared by cooking. The calculated mean levels of gamma-HCH residues in the total diet were 0.004 mg/kg in 1966–67, 0.0035 in 1970–71, 0.003 in 1974–75, 0.0025 in 1975–77, 0.002 in 1979–80, 0.0015 in 1981, and 0.0005 in 1984–85, resulting in dietary intakes of 6.6, 5.5, 4.4, 3.9, 3.0, 2.0, and 0.5 µg/person per day (Egan & Hubbard, 1975; Steering Group on Food Surveillance, 1982, 1986, 1989).

The average daily intake of gamma-HCH in the USA was estimated on the basis of residues found in 30 market-basket composites collected in 30 cities over the period 1964–80, as shown in Table 6.

Table 6. Average daily intake of gamma-HCH in the USA, 1964-80[a]

Year	gamma-HCH intake (µg/kg body weight per day)
1964-69	0.05
1965-70	0.02-0.07[b]
1973	0.0032
1974	0.0084
1975	0.0031
1976	0.0025
1977	0.0039
1978	0.0024
1979	0.0038
1980	0.0028

[a] From Johnson & Manske, 1976; US Environmental Protection Agency, 1980; Gartrell et al., 1985a
[b] From Duggan & Corneliussen (1972)

Infant foods collected in the United Kingdom in 1985–87 generally contained very low levels of gamma-HCH (range, < 0.002 to < 0.01 mg/kg product) (Steering Group on Food Surveillance, 1989). Residues of gamma-HCH were also measured in food composites from 10 cities of the USA in 1974–75 (Johnson et al., 1979). Levels of 0.008–0.012 mg/kg food were found in diets of six-month-old infants, and 0.001–0.007 mg/kg in the diets of two-year-old toddlers. Similar samples collected in 1976–79 in 10 cities consisted of about 50 items of infant food and 110 items of food for toddlers. The daily intake of gamma-HCH was 0.005 µg/kg body weight for infants and 0.01 for toddlers in 1976, 0.006 and 0.008 in 1977, 0.003 and 0.005 in 1978, and 0.001 and 0.006 in 1979 (Gartrell et al., 1985b).

Environmental levels and human exposure

Total-diet studies conducted by the US Food and Drug Administration before 1982 were based on a 'composite sample approach', regardless of the diet involved. Later studies were based on dietary information obtained through surveys, so that the 'total diet' of the US population could be represented by a relatively small number of food items for a large number of age–sex groups (Gunderson, 1988). The average intake of gamma-HCH in the diet of 14–16-year-old boys (mean body weight, 60 kg), estimated using the more recent methods, is shown in Table 7 (S.I. Shibko, letter to IPCS, dated 29 June 1989). The daily intakes in 1982–84 in different age groups were 0.0019 µg/kg body weight per day for 6–11-month-old children, 0.0079 for two-year-old children, 0.0031 for 14–16-year-old girls, 0.0034 for 14–16-year-old boys, 0.0020 for 25–30-year-old women, 0.0025 for 25–30-year-old men, 0.0016 for 60–65-year-old women, and 0.0018 for 60–65-year-old men (Gunderson, 1988). The concentrations for these eight groups in 1984–86, 1987, and 1988 were < 0.01 µg/day for 6–11-month-old infants, < 0.04 for two-year-old children, and < 0.1 for the other six groups (S.I. Shibko, letter to IPCS, dated 29 June 1989).

Table 7. Average daily intake of gamma-HCH in 14-16 year-old boys in the USA[a]

Year	Intake µg/day	µg/kg body weight per day
1982–84	0.204	0.0034
1984–86	0.078	0.0013
1987	0.108	0.0018
1988	0.084	0.0014

[a] From S.I. Shibko, letter to IPCS, dated 29 June 1989

In total-diet studies in Germany, fruit, potatoes, and other vegetables ready for consumption contained 0.001 mg/kg product (Kampe & Andre, 1980). In 17 food groups in Spain, the gamma-HCH concentration ranged from none detected to 0.019 mg/kg product; the level in fat was up to 0.268 mg/kg. The daily intake amounted to 0.0138 mg/person in 1971–72 (Carrasco et al., 1976). In a total-diet study in the Netherlands in 1977, the average concentration of gamma-HCH in 100 samples was 0.03 mg/kg on a fat basis; the highest level was 0.14 mg/kg (Greve & van Hulst, 1977). In another study in the Netherlands, a mean daily intake of 0.002 mg/person was determined for 1976–78 (De Vos et al., 1984).

Data from Canada, Guatemala, Japan, the United Kingdom, and the USA indicate a very low daily intake of gamma-HCH over the years 1971–84. The median values ranged from 0.01 to 0.05 µg/kg body weight (Gorchev & Jelinek, 1985). The daily intake of lindane in the USSR was calculated from a market-basket survey to be about 0.005 mg/day. Cooking reduced this level by a factor of 4.3 (Sizova & Bogomolova, 1976; see Izmerov, 1983).

5.2.2 Intake with drinking-water and air

Edwards (1981) calculated the daily intake of gamma-HCH with drinking-water to be 0.4 ng per person, assuming a daily consumption of 2 litre of water; the median daily intake via air was also calculated to be 17 ng per person, indicating that only small quantities of gamma-HCH are ingested with water and air.

Guicherit & Schulting (1985) measured the concentration of gamma-HCH in the atmosphere and calculated that the daily average intake of a 70-kg Dutch person by inhalation would be 7.2 ng. Another calculation of the average human intake with air, on the basis of ambient concentrations, was 12 ng/day, which represents about 1% of the total daily intake by all routes. The daily intake of gamma-HCH in the USA was estimated to be 0.002 µg/kg body weight by air and 0.07 µg/kg by the oral route (Hildebrandt et al., 1986).

5.2.3 Concentrations in human samples

The concentrations of gamma-HCH in human samples are a good indication of the total exposure of the general population.

5.2.3.1 Blood

Gamma-HCH was detected at a geometric mean of 0.4 µg/litre (range, 0.1–4.1 µg/litre) in the blood of 49 of 62 people in Louisiana, USA (Selby et al., 1969). In a follow-up study, a geometric mean of 0.4 µg/litre (range, 0.1–6.0 µg/litre was found in 47 out of 53 blood samples from pregnant women. Polishuk et al. (1970) found a mean concentration of 0.4 ± 0.8 µg/litre in the blood of 24 pregnant women and 0.3 ± 0.6 µg/litre in the blood of 23 infants living in Israel. Wassermann et al. (1982) found a mean of 4.3 ± 4.8 µg/litre in serum of 10 women in Israel with a normal pregnancy. In a group of 17 women with an abnormal pregnancy (premature birth), a mean concentration of 15.0 ± 7.2 µg/litre was detected. Bercovici et al. (1983) found a concentration of 8.0 ± 4.5 µg/litre in the serum of seven Israeli women with a normal pregnancy and a mean concentration of 8.5 ± 7.8 µg/litre in 17 women with a 'missed abortion'.

Reiner et al. (1977) found a mean concentration of 4.1 ± 0.6 µg/litre (range, 0.5–15.0) in 23 of 147 serum/plasma samples from people living in a town in Yugoslavia. Similar levels were found in other parts of the country (Krauthacker et al., 1980).

Siddique et al. (1981) found gamma-HCH at a mean concentration of 25.0 ± 16.0 µg/litre (range, 8.0–47.0) in the blood of 15 people in India, and Saxena et al. (1981) found a mean concentration of 19.0 ± 12.4 µg/litre (range, 2.4–135.0) in the blood of 100 pregnant women, aged 18–34 years, in the Indian countryside. Nonvegetarian women had higher blood levels than vegetarian women. Kaphalia & Seth (1983) found blood levels of 12.0 (none detected to 71.0) µg/litre in 48 men (aged 20–40 years), 12.00 (range, 5.0–24.0) µg/litre in 16 women (aged 10–30 years) and 16.0 (range, 3.0–64.0) µg/litre in 16 children (< 10 years) in India.

Eckenhausen et al. (1981) found a geometric mean of 0.9 µg/litre of gamma-HCH (range, < 0.4–3.8) in 28 out of 48 pregnant women in the Netherlands. After delivery, a geometric mean of 0.5 (range, 0.2–19.0) µg/litre was measured in 24 out of 66 women, and a mean of 0.5 (range, < 0.3–34) µg/litre in 33 out of 86 babies.

Blood samples from Dutch citizens were analysed in 1978 (70 samples), 1980 (48 samples), 1981 (127 samples), and 1982 (54 samples); the concentration of gamma-HCH was in the range < 0.1–0.2 µg/litre blood (Greve & van Harten, 1983; Greve & Wegman, 1985). Blok et al. (1984) measured the levels of gamma-HCH in the blood of 65 healthy volunteers (34 women and 31 men) in the Netherlands and found residues in approximately two-thirds of the people studied. The median concentration in both men and women was 0.2 µg/litre (range, none detected to 0.6 µg/litre). Bertram et al. (1980) found a median concentration of 1.18 µg/litre (range, none detected to 2.94) in 118 whole-blood samples in Germany.

In 8 of 35 serum samples from mothers in Norway and in 6 of 35 corresponding samples of umbilical cord serum, the levels of gamma-HCH ranged from 0.2 to 27 µg/kg wet weight. In serum samples from immigrant mothers and in 5 of 7 corresponding umbilical cord serum samples, the levels were 0.1–3.4 µg/kg wet weight (Skaare et al., 1988).

5.2.3.2 Adipose tissue

Mes et al. (1982) analysed 99 samples of adipose tissue from autopsied accident victims in different areas of Canada and found an average concentration of gamma-HCH of 0.003 mg/kg (wet weight) (range, 0.001–0.03 mg/kg). Nearly all of the samples (90%) contained gamma-HCH.

In 567 samples of adipose tissues from Dutch citizens analysed for the presence of gamma-HCH in 1968–83, the concentration varied from < 0.01 to 0.04 mg/kg; the highest levels were found for 1968–76 (Greve & van Harten, 1983; Greve & Wegman, 1985). Bertram et al. (1980) found a median concentration of 0.05 mg/kg (range, none detected to 0.44) in 72 samples of adipose tissue from people in Germany. In specimens of subcutaneous fat taken in 1982–83 from 48 children (34 < 1 year; 14 aged 2 years), the concentration of gamma-HCH was 0.04 mg/kg fat (range, 0.01–0.21 mg/kg). The average concentration was highest in infants aged 0–6 weeks, at 0.07 mg/kg fat (range, 0.02–0.21 mg/kg) (Niessen et al., 1984). The results of nine studies carried out in Germany in 1969–83 (598 samples) gave a mean concentration for gamma-HCH of 0.01–0.34 mg/kg on a fat basis (Hildebrandt et al., 1986).

Twenty-nine samples of adipose tissue were taken at necropsy and 24 at surgery in the Poznan district in Poland and compared with 100 samples from residents of the Warsaw region. The mean concentration of gamma-HCH in Poznan was 0.020 $\pm$ 0.018 and that in Warsaw, 0.022 $\pm$ 0.003 mg/kg (Szymczynski et al., 1986).

The mean concentration of gamma-HCH in 360 samples of adipose tissue collected in eight regions of Japan in 1974 was 0.035 mg/kg (Takabatake, 1978).

5.2.3.3 Breast milk

Breast milk is a major route of elimination of organochlorine pesticides and polychlorinated biphenyls in women.

In a Swedish study, the levels of gamma-HCH in mothers' milk were found to be related to their dietary habits: levels in lacto-vegetarians were lower than those in subjects who ate a mixed diet, and these were lower than those found in mothers who ate a mixed diet which regularly included fatty fish from the Baltic Sea (Noren, 1983).

A significant correlation was found between the concentration of gamma-HCH in breast milk and the amounts of meat products and animal fat in the diet. In addition, concentrations of gamma-HCH in breast milk appeared to be higher in rural areas than in urban areas (Cetinkaya et al., 1984).

Variations in residue levels in human milk during a lactation period of up to 9 months were investigated in five women aged 23–36 years in Germany: gamma-HCH concentrations were 0.004–0.022 mg/kg on a fat basis, and no essential change in residue level occurred over the lactation period (Fooken & Butte, 1987).

More than 7100 samples of breast milk were analysed in Germany between 1969 and 1984 by 20 authors, and the results were summarized by Hildebrandt et al. (1986). The mean concentration of gamma-HCH was 0.01–0.11 mg/kg on a fat basis, but a mean concentration of 0.45 mg/kg was found for a group of 137 samples. A slow decrease in the mean concentration was observed between 1978 and 1984. The average concentration in human milk (2709 samples) in Germany in 1979–81 was 0.06 mg/kg on a fat basis (Fooken & Butte, 1987); in 1981–83, the average level in 132 samples of breast milk was 0.032 mg/kg milk fat (Cetinkaya et al., 1984). The results of other studies were summarized by the Deutsche Forschunsgemeinschaft (1978, 1983). The results for other countries are comparable to those for Germany, although higher levels (mean, 0.33 mg/kg) were measured in Czechoslovakia in 1971–73 (Deutsche Forschungsgemeinschaft, 1983). Average concentrations in human milk in France between 1970 and 1975 were in the order of 0.06–0.07 mg/kg (fat basis) (Rhône-Poulenc Agrochimie, 1986).

Tuinstra (1971) analysed 40 breast milk samples from young mothers (18–32 years of age) in the Netherlands and found a median concentration of 0.01 mg/kg on a fat basis (range, none detected to 0.04 mg/kg on a fat basis). The median concentration of gamma-HCH in 278 samples of human milk collected in 11 maternity centres in the Netherlands was < 0.01 mg/kg on a fat basis; the highest value found was 0.08 mg/kg (Greve & Wegman, 1985).

Mes et al. (1986) studied 210 breast milk samples from five regions across Canada and found gamma-HCH at a mean concentration of 5 µg/kg (on a fat basis). Davies & Mes (1987) studied 18 breast milk samples from Canadian, Indian, and Inuit mothers in Canada whose fish consumption was comparable to the national rate. The level of gamma-HCH was 7 µg/kg in milk fat of the indigenous population, in comparison with 5 µg/kg in a national survey.

Vukavic et al. (1986) measured gamma-HCH in 59 samples of colostrum collected in Autumn 1982 (26 samples) and Spring 1983 (33 samples) in Yugoslavia from healthy nursing mothers on the third day after delivery. The concentrations of gamma-HCH were significantly higher in Autumn (1.71 ± 0.44 µg/litre) than in Spring (0.67 ± 0.12 µg/litre).

Breast milk samples from four women in Iraq, examined once a week for 20 weeks, contained average levels of 0.009, 0.005, 0.134, and 0.005 mg/kg whole milk. Gamma-HCH levels in placental tissue from these four donors were 0.004, 0.011, 0.013, and 0.006 mg/kg, respectively. Fluctuations in the residue levels were seen to be due to variations in the daily dietary intake and variations in the fat content of the breast milk (Al-Omar et al., 1986).

6. KINETICS AND METABOLISM

6.1 Absorption

6.1.1 Oral administration—experimental animals

The uptake of lindane by rats or mice has been studied after oral administration. Direct information on the velocity of uptake from the gastrointestinal tract is available, which can be supplemented by information from studies in which excretion of orally administered radioactive lindane was followed.

Lindane taken up from the intestines is transferred almost exclusively to the blood. No significant amount was found in the lymph of rats after injection of 0.05 or 0.1 µmol into the loops of the small intestines *in vivo*. Absorption was rapid: 29–53% of the injected material was absorbed from the intestinal loops within the first 30 min (Turner & Shanks, 1980). Uptake of lindane from the intestines of rats given 12.5 mg in oil over five days was less effective in animals depleted of their intestinal microorganisms by maintenance under aseptic conditions than in conventional rats. The asepticized rats also excreted more unchanged lindane in the faeces than conventional animals (Macholz et al., 1983).

6.1.2 Dermal application—experimental animals

Hawkins & Reifenrath (1984) developed an apparatus to determine the evaporation and percutaneous penetration of hexachloro-[U-^{14}C]-cyclohexane (lindane) *in vitro*, applying a dose of 4 µg/cm^2 on pig skin. Evaporation accounted for 26 ± 5%; skin oxidation for 43 ± 17%; and percutaneous penetration for only 0.7 ± 0.3% of the applied radiolabel. Reifenrath et al. (1984) also evaluated models consisting of human or pig skin grafted onto the congenitally athymic nude mouse, hairless dogs and weanling Yorkshire pigs for predicting skin penetration in humans. A radioactive dose of 0.05 µCi of [U-^{14}C]-labelled lindane (98%) was applied topically to 1.27 cm^2 (4 µg/cm^2) of each model, and radiolabel (percentage of applied dose) was measured in urine and faeces, skin scrub, application site, and carcass. Incomplete excretion of the label following topical application was corrected for by parental (subcutaneous) administration of 2 µCi in propylene glycol. The results showed significant correlations between the values for human skin and those for human skin grafted on athymic mice and for weanling Yorkshire pigs, but no correlation was found between the values for humans and those for the hairless dog or for pig skin grafted on athymic nude mice.

Dermal absorption has also been investigated in rats and rabbits. Groups of 24 male Charles-River Crl:CD(SD)BR rats and male Hra:(NZW)SPF rabbits were given a single dermal application of lindane (20% emulsifiable concentrate to which ^{14}C-lindane had been added) on clipped skin at doses of 0.1, 1, or 10 mg/rat and 0.5, 5, and 50 mg/rabbit, corresponding to 0.02, 0.2, and 2 mg/cm^2, respectively. Urine and faeces were collected 0.5, 1, 2, 4, 10, or 24 h after application from four animals per dose level; and four animals per group were bled and sacrificed 0.5, 2, 4, 10, and 24 h after application of the test material. The ratio between the quantity absorbed at a dose of 1 mg and that absorbed at 10 mg, as well as that between 0.1 and 1 mg/rat at equal exposure duration, decreased with increasing exposure time and concentration. On a group basis, the total recoveries were 75–85% for rats and 75–82% for rabbits. A significant fraction of the applied dose was found in the urine: 16, 15, and 12%, respectively, at application levels of 0.1, 1, and 10 mg/rat. The corresponding values for rabbits were 46, 29, and 41% for doses of 0.5, 5, and 50 mg/rabbit. Much lower levels were found in the faeces. Total adsorption (24 h) increased in rats from 5% of the applied amount at the highest dose to 28% at the lowest exposure. For rabbits, penetration was more rapid, and adsorption at 24 h ranged from 17 to 56% of the applied dose. The applied doses per unit surface area were approximately the same, permitting a comparison of penetration rates. The average penetration rates after 24 h for rats were 0.2, 2, and 4 µg/cm^2 per hour for groups treated with 0.02, 0.2, and 2 mg/cm^2. The penetration rates after 24 h for rabbits were 0.5, 3, and 14 µg/cm^2 per hour for the groups treated with 0.02, 0.2, and 2 mg/cm^2. These studies indicate that appreciable absorption of lindane takes place after dermal application (Bosch, 1987a, b).

1.1.3 Other routes—experimental animals

When doses of 40 or 80 mg/kg body weight of a mixture of ^{14}C- and ^{36}Cl-gamma-HCH in rapeseed oil were injected intraperitoneally into rats, 25% was absorbed within 1 h and at least 90% after 1 day. Four days after the injection, only traces of lindane were left in the abdominal cavity. One day after the injection, about 40% of the applied dose was found in the organs and tissues (Koransky et al., 1963).

Kinetics and metabolism

6.2 Distribution

6.2.1 *Oral administration—experimental animals*

After uptake, lindane is distributed to all organs and tissues in the body of laboratory animals, at measurable concentrations within a few hours.

When lindane was administered orally to rats at doses of 1, 10, or 100 mg/kg diet for up to 56 days, the highest concentrations were found in adipose tissue. The fat:blood ratio in this study was very close to 150 at all times, whereas the liver:blood ratio was 3.4–3.5. Lindane concentrations in organs reached a maximum after 2–3 weeks and slowly decreased thereafter. The authors did not differentiate between males and females (Oshiba, 1972).

Twenty-four hours after oral administration of ^{14}C-labelled lindane at 8 mg/kg body weight in sunflower oil to rats for 10 days, more than 35% of the administered activity was deposited in fat. Muscle and kidneys contained 3.5 and 3.7%, respectively, and all other organs that were analysed contained less than 1%; 17.4% was found in urine and 13.8% in faeces. In total, only 78.7% was recovered; however, skin was not analysed in this study. (Other studies have demonstrated that a significant amount of lindane administered to rats and mice is deposited in the skin, so that some of the missing 21.3% of the total applied dose could have been there.) After an additional 48 h, the concentrations in all organs were reduced to about one-half of the values seen after 24 h, showing that no single organ retained lindane significantly longer than the others. Urine contained 24.5% and faeces, 20.9% (Seidler et al., 1975).

After single oral doses to rats, the fat:blood ratio ranged between 145 and 206 and the brain:blood ratio between 4 and 6.5 (Vohland et al., 1981).

After continuous dietary administration of lindane at doses of 0.2, 0.8, 4.0, 20, or 100 mg/kg diet for 13 weeks to Wistar KFM-Han rats, the highest concentrations were reached in the fatty tissue. At the highest dose, the fat:blood ratio was 44 in males and 69 in females and the liver:blood ratio was 5.3 in males and 9.6 in females. After six weeks with no further administration, lindane concentrations in organs were close to the control values (Suter et al., 1983).

The distribution of lindane in brain after oral administration at 30 mg/kg or intravenous administration at 0.3 mg/kg was studied using autoradiography–imaging analysis and dissection–liquid scintillation counting techniques. The two routes of administration gave similar results. A heterogeneous distribution of label in brain regions was observed: the radiolabel concentration in the white matter was higher than that in thalamus, mid-brain, pons and medulla at different times relative to the mean value for whole brain. The affinity of lindane for white matter and myelinated structures was related to its lipophilic behaviour (Sanfeliu et al., 1988).

Mosha et al. (1986) studied the distribution and elimination of gamma-HCH in adult female goats. Eight goats were administered a daily dose of 6 mg/kg body weight by gavage for five consecutive days. Blood and milk were collected before exposure and during 10–60 days after exposure, and organs and tissues were collected and analysed. One goat was used as a control. The blood concentration of gamma-HCH was approximately 0.1 mg/litre during the dosing period and decreased gradually thereafter; none was detectable after day 20. The concentrations in milk were about eight times higher than those in blood but decreased in parallel. The concentration in fat samples on day 7 was 1.4 mg/kg, but those in other tissues were about 0.1 mg/kg.

5.2.2 Inhalation—experimental animals

After rats were exposed to lindane at doses of 0.02, 0.1, 0.5, or 5 mg/m^3 for 90 days by inhalation, the highest concentrations were found in fatty tissues. At 5 mg/m^3, the fat:serum ratio was 150 in males and 464 in females; at 0.5 mg/kg, it was 161 in males and 245 in males; at 0.1 mg/kg, 137 in males and 429 in females; and at 0.02 mg/kg, 92 in males and 377 in females. At the 5 mg/m^3 exposure level, the liver:blood ratios were 1.9 in males and 4.2 in females and the brain:serum ratios were 9 in males and 23 in females. These values suggest that higher concentrations are reached in fat after inhalation than after oral administration, whereas concentrations in liver appear to be somewhat lower after inhalation. After a recovery period of four weeks, concentrations in all organs had decreased to the control values (Oldiges et al., 1983).

5.2.3 Other routes

One day after intraperitoneal injection to rats of a mixture of ^{14}C- and ^{36}Cl-lindane in rapeseed oil, the highest contents were those of skin and fat—15.7% and 10.7%, respectively. Less than 1% was found in all other organs, including the central nervous system (Koransky et al., 1963). When lindane or deuterated lindane was administered intraperitoneally to rats at a dose of 10 mg/kg body weight, about 40 mg/kg fat were found after one day in both males and females. At that time, the blood concentration in males was 0.2 mg/litre; 1–2 mg/kg were present in brain and 0.7 mg/kg in skeletal muscle. Deuterated lindane was found at 110 mg/kg in depot fat of males, and the levels in brain and muscle were about twice those of undeuterated lindane (Stein et al., 1980).

Mottram et al. (1983) studied the metabolic fate of lindane in three groups of two white female pigs. The pigs were sprayed once with either 5.6 g or 1.4 g of an anti-louse spray, which represented 16 and 4 times, respectively, the normal dose of 350 mg/pig. Five animals served as controls. Rapid accumulation of lindane occurred, and several metabolites were found in adipose tissue. The main metabolite was 1,2,4-trichlorobenzene. The residues were eliminated rapidly from adipose tissue, so that 30 days after treatment, the residual concentration in pigs sprayed with a dose 16 times the recommended rate was no greater than that in the untreated controls.

Residue levels were also investigated in four lactating goats following oral and topical application of labelled lindane (Wilkes et al., 1987a,b). Two Alpine goats were housed in metabolic cages and administered lindane (purity not stated), spiked with ^{14}C-lindane, in the diet at doses of 1 and 10 mg/kg twice daily for four days; they were sacrificed 12–14 h after the last dose. One Alpine goat received two topical applications at a seven-day interval of a lindane solution (purity unspecified) containing 11.0 mg/ml to a shaved area that represented about 25% of the body surface area, to simulate total body spray; it was sacrificed 48 h after the last application. A Nubian goat was not only shaved over the same extent but also had its remaining hair clipped to approximately 3 cm, and the whole surface of the animal was treated with lindane, to simulate dip treatment; this animal was sacrificed 24 h later. After sacrifice, radiolabel was measured in tissues and in the intestinal contents; radiolabel in exhaled carbon dioxide was measured on one occasion.

Total recovery of radiolabel was low: approximately 50% in the study by oral administration and 16–30% in the study by dermal application. To determine the reason for the losses after oral administration, the fate of labelled lindane in rumen fluid at 37 °C was investigated *in vitro*. About 55% of the radiolabel was recovered as ^{14}CO$_2$, and 38% remained in the rumen fluid. The authors stated that their results "clearly show that volatile ^{14}C-labelled organics were evolved from ^{14}C-lindane fortified rumen fluid," and "that the losses in the *in vivo* studies were due to volatile ^{14}C-lindane metabolites." Attempts to trap the labelled volatile compounds, however, proved unsuccessful.

From 35 to 46% of the radiolabel administered orally was excreted in the urine over a period of 4 days, and 10–12% of the dermally applied dose was found in urine over 8–9 days. Much lower activities were found in faeces and insignificant amounts in expired air. The amount of radiolabel in whole milk after oral administration reached a plateau after 2–3 days, corresponding to a total concentration of 0.4 ppm (6–8 ppm in the fat) at the lower dose (2 mg/kg per day) and 3 ppm (about 50 ppm in the fat) at the higher exposure level (20 mg/kg per day). Significant activity was also found in the milk after dermal administration, corresponding to levels of 0.1–0.7 ppm in whole milk.

5.3 Metabolic transformation

The metabolism of lindane is initiated by one of four possible reactions:
- Dehydrogenation leads to the formation of gamma-HCH;
- Dehydrochlorination leads to the formation of gamma-PCCH;
- Dechlorination leads to the formation of gamma-tetrachlorohexene;
- Hydroxylation leads to the formation of hexachlorocyclohexanol.

These compounds must be considered as intermediates, and the initial reactions are followed by a series of further dehydrogenating, dechlorinating, dehydrochlorinating, and hydroxylating steps.

A large number of metabolites and end-products occur during the metabolism of lindane (see section 6.3.2). Detailed descriptions and schemes of the metabolic pathway of lindane leading to the various isomeric metabolites have been proposed (Engst et al., 1970, 1977, 1978a, 1979a; Kurihara & Nakajima, 1974). These pathways involve metabolites that have not as yet been detected. The missing metabolites may be very unstable compounds which are rapidly transformed to other intermediates and thus escape detection. Another possibility is that conjugates are formed which are tightly bound to proteins *via* sulfur, and these could be detected only after the complex is hydrolysed.

The essential steps in the metabolism of lindane are known, and these are shown, with the main metabolites, in Figure 2 and are discussed below.

Kinetics and metabolism

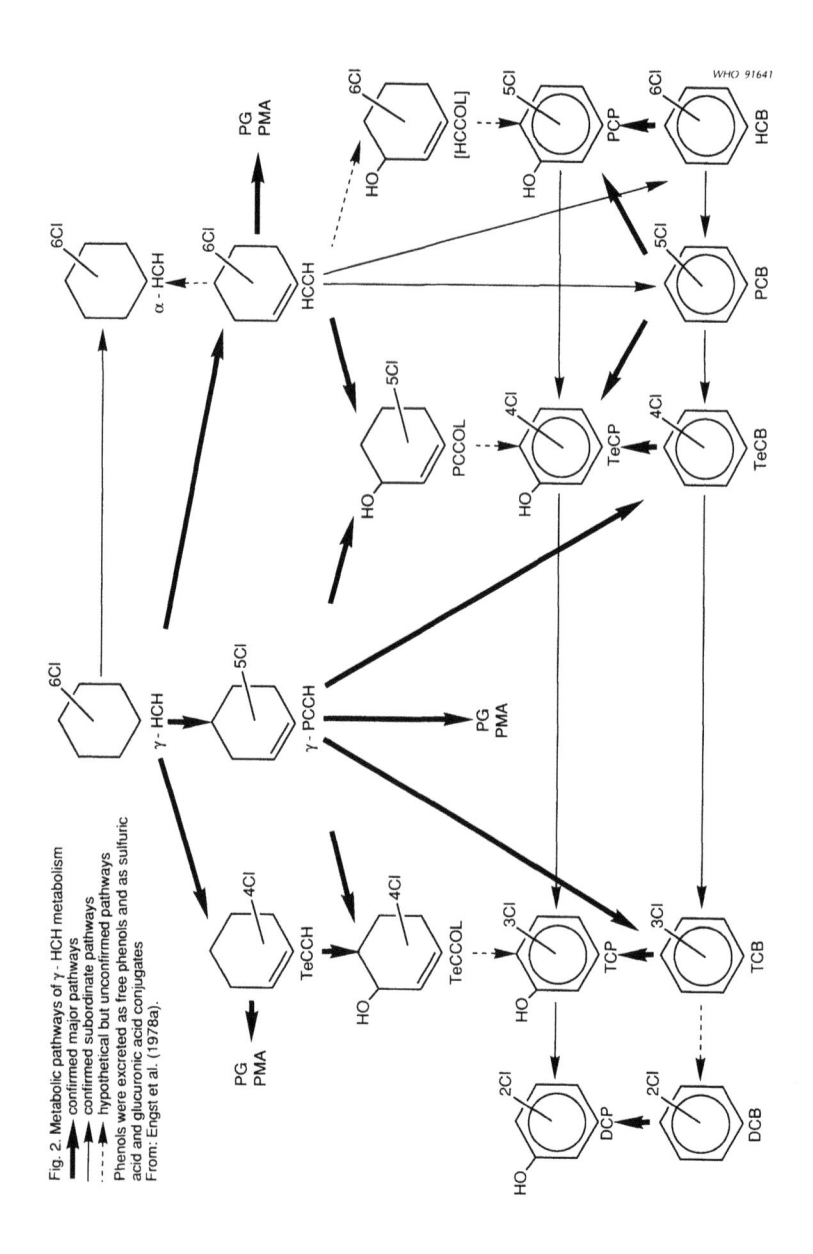

Fig. 2. Metabolic pathways of γ-HCH metabolism
⬆ confirmed major pathways
⬆ confirmed subordinate pathways
---- hypothetical but unconfirmed pathways
Phenols were excreted as free phenols and as sulfuric acid and glucuronic acid conjugates
From: Engst et al. (1978a).

6.3.1 Enzymatic involvement

Lindane is converted by enzymatic reactions, mainly in the liver. One group of enzymes involved in the biotransformation of lindane is microsomal, e.g., cytochrome P-450-dependent monooxygenases. Five groups of male Wistar rats were injected intraperitoneally with gamma-HCH at 25 mg/kg body weight on four consecutive days to investigate the induction of cytochrome P-450 in liver microsomes. Gamma-HCH was found to be a 'mixed type' inducer which mediates the induction of cytochrome P-450 b/e, c and d forms (Kumar & Dwivedi, 1988). These enzymes are involved in hydroxylation, dehydrogenation, and dechlorination. Other hepatic cytosolic enzymes are involved in the dehydrochlorination reaction. The intermediate metabolites or end-products of the biotransformation may result as a consequence of the four enzymatic reactions listed above.

A cytochrome P-450-dependent dehydrogenation reaction was described in rat liver microsomes *in vitro*. Incubation of lindane with rat liver homogenates resulted in the formation of hexachlorocyclohex-1-ene, and the authors proposed that this dehydrogenation is an important initial step in the metabolism leading to the detoxication of lindane (Chadwick et al., 1975).

Stein et al. (1977) found that at least two independent pathways were involved in the metabolism of lindane. The first is the possible formation of unstable intermediates, such as hexachlorocyclohexanol, after an initial hydroxylation leading to the main metabolite, 2,4,6-trichlorophenol (2,4,6-TCP), and involving cytochrome P-450. The second pathway includes dehydrogenation of lindane to 1,2,3,4,5,6-HCCH, subsequent hydroxylation and dehydrochlorination to 2,3,4,6-tetrachlorophenol. TCP and tetrachlorophenol are formed *in vitro* in a ratio of about 2:1.

In the presence of oxygen and NADPH *in vitro*, rat liver microsomes metabolize lindane mainly to 2,4,6-TCP (Tanaka et al., 1977, 1979). 3,4,5,6-Tetrachlorocyclohexene was identified as an intermediate after lindane was incubated under N_2 with liver preparations (Chadwick et al., 1978; Kurihara et al., 1979). Experiments with rat liver preparations *in vitro* demonstrated the importance of glutathione in the metabolism of lindane. Glutathione enhanced the conversion of lindane to dichlorophenol (DCP) by a factor of 3–4, but conjugates were formed only in the presence of liver cytosol protein as a source of glutathione transferase. The initial step appears to be dehydrochlorination to 1,3,4,5,6-PCCH, followed by conjugation and further dehydrochlorination to DCP—mainly 2,4-DCP. The DCPs found in the urine of rats are a mixture of different isomers. The rate of formation of *S*-(dichlorophenyl) glutathione from HCH in rat liver cytosol apparently depends on gradual monodehydrochlorination, and the enzymatic transfer of reduced L-glutathione onto PCCH is not preceded by a second dehydrochlorination (Stein et al., 1977; Portig et al., 1979; Tanaka et al., 1979).

When a 1:1 mixture of lindane and the corresponding hexadeuterated compound was fed to Wistar rats, the ratio of monochlorophenol, DCP, and TCP, excreted as mercapturic acids, showed an isotopic effect. The rate-limiting step for the formation of DCP and TCP is either a dehydrochlorination or a dehydrogenation of lindane, whereas formation of monochlorophenol must be initiated by dechlorination to tetrachlorocyclohexane, followed by conjugation with glutathione (Kurihara et al., 1980).

Lindane was converted mainly to gamma-HCCH by rat liver microsomes, and significant amounts of 2,4,6-TCP and 2,3,4,6-tetrachlorophenol were detected. Human liver microsomes converted lindane into four major metabolites: gamma-1,2,3,4,5,6-hexachlorocyclohex-1-ene; gamma-1,3,4,5,6-PCCH; beta-1,3,4,5,6-PCCH, and 2,4,6-TCP. Smaller amounts of 2,3,4,6-tetrachlorophenol and pentachlorobenzene were found (Fitzloff et al., 1982).

Human and rat liver microsomes converted the lindane metabolites gamma-PCCH and 3,4,6/5-PCCH to 1,2,4-trichlorobenzene, 1,2,3,4-tetrachlorobenzene, 2,4,5-TCP, 3,4,5/6-pentachloro-2-cyclohexen-1-ol and beta-PCCH-oxide or 3,4,6/5-PCCH-oxide. The identity of the beta-PCCH-oxide was confirmed by column chromatography and gas-liquid chromatography–mass spectrometry. It is stable to hydrolysis by microsomal epoxide hydrolase (E.C.3.3.2.3.) and under various aqueous acid conditions. Its toxicological role is still unknown. Although this compound is structurally related to epichlorhydrin and epoxides, it was not mutagenic to *Salmonella typhimurium* strain TM677 (Fitzloff & Pan, 1984).

Two groups have reported the formation of trace amounts of chlorinated benzenes from lindane in rats. Hexachlorobenzene was found in faeces (Gopalaswamy & Aiyar, 1984) and pentachlorobenzene in brain (Vohland et al., 1981). The two studies are consistent in so far as the identified amounts of chlorobenzenes are extremely low and near the detection limit; however, they are also contradictory, because hexachlorobenzene was found exclusively in the first study and pentachlorobenzene in the second. It is impossible to clarify whether artefacts were measured in these studies, e.g., enrichment of impurities in the starting material. If indeed chlorinated benzenes are formed from lindane, the amounts obtained are insignificant compared to those of other metabolites.

7.3.2 Identification of metabolites

Metabolites of lindane have been identified in a large number of studies, *in vivo* in body fluids, urine, faeces, organs, and tissues, and *in vitro*. Most of the *in-vivo* studies were carried out with rats, but similar results were obtained in other animal species.

The following metabolites have been identified:

Cycloalkenes: 1,2,3,4,5,6-hexachlorocyclohexene (HCCH); 1,3,4,5,6-pentachlorocyclohexene (PCCH); 3,4,5,6-tetrachlorocyclohexene; 2,3,4,5,6-pentachloro-2-cyclohexen-1-ol; and 2,3,4,6- and 2,4,5,6-tetrachloro-2-cyclohexene-1-ol (Chadwick & Freal, 1972a; Freal & Chadwick, 1973; Chadwick et al., 1975; Engst et al., 1976; Kujawa et al., 1977; Tanaka et al., 1977; Chadwick et al., 1978; Engst et al., 1978b; Kurihara et al., 1979; Vohland et al., 1981; Fitzloff et al., 1982; Mottram et al., 1983; Sanfeliu et al., 1988).

Chlorobenzenes: 1,2,3,4,5,6-hexachlorobenzene, 1,2,3,4,5-pentachlorobenzene and 1,2,3,5-, 1,2,4,5-, and 1,2,3,4-tetrachlorobenzene (Aiyar, 1980; Vohland et al., 1981; Gopalaswamy & Aiyar, 1984; Artigas et al., 1988). Mono-, di-, tri-, and tetrachlorobenzenes have been reported in pigs (Mottram et al., 1983).

Chlorophenols: 2,3,4,5,6-pentachlorophenol; 2,3,4,5-, 2,3,4,6-, and 2,3,5,6-tetrachlorophenol; 2,3,5-, 2,4,5-, and 2,4,6-TCP; and 2,4- and 3,4-DCP (Grover & Sims, 1965; Chadwick & Freal, 1972a,b; Freal & Chadwick, 1973; Kurihara & Nakajima, 1974; Chadwick et al., 1975; Engst et al., 1976; Kujawa et al., 1977; Stein et al., 1977; Tanaka et al., 1977; Engst et al., 1978b; Tanaka et al., 1979; Aiyar, 1980; Fitzloff et al., 1982).

Conjugates of these compounds:

– 2,3- and 2,6-DCP were conjugated with *glutathione* (Portig et al., 1979);
– 2,4-DCP; 2,3,5-, 2,4,5-, and 2,4,6-TCP; 2,3,4,5-, 2,3,4,6-, and 2,3,5,6-tetrachlorophenol; 2,3,4,5,6-PCP and tetrachloro-2-cyclohexene-1-ol were conjugated with *glucuronic acid* (Grover & Sims, 1965; Kurihara & Nakajima, 1974; Engst et al., 1976; Chadwick et al., 1981);

– 2,3,5-, 2,4,5-, and 2,4,6-TCP; 2,3,4,5- and 2,3,4,6-tetrachlorophenol; 2,3,4,6- tetrachloro-2-cyclohexene-1-ol; and 2,3,4,5,6-pentachloro-2-

cyclohexen-1-ol were conjugated with sulfate (Grover & Sims, 1965; Kurihara & Nakajima, 1974; Chadwick et al., 1981); and

– 4-monochlorophenol; 2,4- and 3,4-DCP and 2,3,5- and 2,4,5-TCP were conjugated with mercapturic acids (Grover & Sims, 1965; Kurihara et al., 1979, 1980).

6.3.3 Metabolites identified in humans

Engst et al. (1978b) analysed urine from workers apparently exposed to technical-grade HCH (during manufacture?) and found alpha-, beta-, gamma-, and delta-HCH, traces of hexa- and pentachlorobenzene, gamma- and delta-PCCH, pentachlorophenol, 2,3,4,5-, 2,3,4,6-, and 2,3,5,6-tetrachlorophenol, several TCPs, as well as glucuronides of some of these metabolites. The PCCHs, tetrachlorophenol, hexachlorobenzene, and pentachlorophenol were also identified in blood.

The urine of 21 men working in the production of gamma-HCH with a purity of 99.8% from technical-grade HCH (16% alpha-, 7% beta-, and 45% gamma-HCH) was examined for the presence of chlorinated phenols. External and internal exposure was estimated from measurements of the concentrations of HCH isomers in the air of the workroom and in serum samples. The men had been employed for periods ranging from a few months up to 30 years (mean, 10.6 years), and they were aged 24–62 years (mean, 46 years). Fourteen mono-, di-, tri-, and tetrachlorophenols and seven dihydroxychlorobenzenes of unknown configuration were identified in urine. The main metabolites were 2,4,6-, 2,3,5-, and 2,4,5-TCP, which were excreted in nearly equal quantities. The mean concentrations of alpha-, beta-, and gamma-HCH in the serum of exposed workers were 49 (range, 11–138), 82 (17–434) and 52 (9–188) µg/litre; the levels in controls were < 1.0 µg/litre. The air concentrations of alpha-, beta-, and gamma-HCH were 2–4, 1–3, and 23–63 µg/m^3 (Angerer et al., 1983).

6.4 Elimination and excretion in expired air, faeces, and urine

6.4.1 Oral administration

In mammals, including human beings, lindane is excreted very rapidly in urine and faeces after metabolic degradation; only small quantities are eliminated unchanged (Seidler et al., 1975). As lindane is subjected to four types of reaction—dehydrochlorination, dechlorination, dehydrogenation, and oxidation—many intermediate metabolites are found, the nature of which depends on the initial reactions. Nevertheless, the excreted metabolites are all various isomers of DCP, TCP, and tetrachlorophenol, which are excreted either free or in a conjugated form with glucuronic or sulfuric acid or *N*-acetylcysteine (Rhône-Poulenc Agrochimie, 1986).

6.4.1.1 Rat

Sprague–Dawley rats were fed diets containing lindane at 400 mg/kg diet for 5 weeks. Within 24 h, mainly 2,3,4,6- and 2,3,4,5-tetrachlorophenols, 2,3,5-, 2,4,5-, and 2,4,6-TCPs and 3,4-DCP were found in the free form in urine and faeces, at 27.1%, 4.3%, 8.4%, 14.7%, and 51.1%, respectively (Chadwick & Freal, 1972a; Chadwick et al., 1975).

After ^{14}C-lindane was administered orally to rats at 8 mg/kg body weight for 10 days, 23% of the metabolites in faeces and urine were in the free form, and 77% in conjugated form, partly as glucuronides (Seidler et al., 1975).

Metabolites were extracted from the urine of male Wistar rats that had received 19 daily oral doses of 8 mg/kg body weight. After hydrolysis of conjugates, the metabolites found were 2,4,6-TCP, 2,3,4,6- and 2,3,5,6-tetrachlorophenol, tetrachlorocyclohexenol, pentachlorocyclohexenol, and pentachlorophenol (Engst et al., 1976). Formation of 2,4,5,6- and 2,3,4,6-tetrachlorocyclohexenol was confirmed by Chadwick et al. (1978) in a study with Sprague–Dawley rats fed diets containing lindane at 400 mg/kg diet for one month.

6.4.1.2 Rabbit

Five rabbits fed gelatin capsules containing ^{14}C-lindane at 3–12 mg per animal twice weekly for 26 weeks excreted 54% of the radiolabel in urine and 13% in faeces. About 56% of the urinary metabolites were soluble; those that were identified were 2,3,5-, 2,4,5-, and 2,4,6-TCP, 2,3,4,6-tetrachlorophenol, 2,3- and 2,4-DCP and 2,3,4,5-tetrachlorophenol. The presence of seven chlorophenols and six chlorobenzenes was indicated (Karapally et al., 1973).

6.4.2 Other routes

6.4.2.1 Mouse

Lindane metabolites were analysed in the urine of mice after a single intraperitoneal injection of ^{14}C-lindane at 16 or 21 µg/mouse. Within 3 days, 57% of the total radiolabel had been excreted in urine, mainly as conjugates with glucuronic or sulfuric acid. About 25% of the excreted conversion products were 2,4,6-TCP, and 4–6% was 2,4-DCP; 41–46% of the chlorophenols were conjugated and 3% in the free form (Kurihara & Nakajima, 1974).

6.4.2.2 Rat

Intraperitoneal administration to rats of lindane in arachis oil at daily doses of 40 mg/kg body weight (total, 4 g) was followed by urinary excretion of 2,3,5- and 2,4,5-TCP, either in free form or as sulfuric and glucuronic acid conjugates (Grover & Sims, 1965).

One day after intraperitoneal injection of a mixture of ^{14}C- and ^{36}Cl-labelled lindane to rats, 18.97% of the radiolabel was found in the excreta; 7.39% was still not absorbed, indicating again that elimination of lindane begins during the absorption phase. After 4 days, 52% of the total activity was found in the excreta. The resulting half-time was about 4 days (Koransky et al., 1963). An even shorter half-time, of 1–2 days, was seen in depot fat in another study after intraperitoneal injection of 10 mg/kg to rats (Stein et al., 1980).

After intraperitoneal administration of 40 mg/kg body weight to rats, 20% of the total dose left the body *via* the faeces and 80% *via* the urine (Koransky et al., 1963, 1964). In another study, however, the amounts of radiolabel excreted by rats in urine and faeces were about equal (Seidler et al., 1975). Only traces of unchanged lindane were found in faeces and urine. Of the chlorine derived from lindane that is excreted in the urine, about 60% is inorganic and 40% is organic (Koransky et al., 1964).

Mono-, di-, and trichlorophenyl mercapturic acids were found to be the main metabolites after intraperitoneal administration of lindane at 17.2 and 34.4 µmol to male Wistar rats, accounting for more than 60% of the urinary metabolites (Kurihara et al., 1979).

6.4.2.3 Human

The urinary excretion of radiolabel after intravenous administration of ^{14}C-labelled lindane to six human subjects at 1 µCi in propylene glycol was 24.6% ± 6.1 of the administered dose within five days. About 80% was excreted in the first 24 h. The half-time was 26 h (Feldmann & Maibach, 1974).

6.5 Retention and turnover (experimental animals)

The highest concentrations of gamma-HCH in the bodies of mice were found 3 h after oral administration of a single dose of 1.2 mg/mouse. After 72 h, 270 µg of the original 1200 µg/animal remained. A half-time of 2–3 days can be deduced from these results (van Asperen, 1958).

When lindane was fed to rats for 56 days at doses of 1, 10, or 100 mg/kg diet, organ contents increased to a maximum within 2–3 weeks, depending on the dose. From that time on, the concentrations in all organs decreased slowly, and equilibria were approached by the end of the administration period. The concentration ratios between different organs and blood remained constant throughout the time of administration, and when treatment was stopped the concentrations in all organs, including adipose tissue, decreased rapidly. Similar results were seen after starvation for 6 days, whereas diets rich in fat or protein accelerated the reduction of the lindane content in organs and tissues. The kinetics of excretion indicate a half-time of 3–4 days for oral administration of a single dose of ^{14}C-lindane (Oshiba, 1972).

After continuous feeding of lindane in corn oil at 50 mg/kg diet for 60 days to Osborne–Mendel rats, a constant equilibrium concentration of about 50 mg/kg was reached in adipose tissue within 9 days. After cessation of lindane administration, the concentration dropped to values between 2.5 and 11.5 mg/kg tissue within 9 days (Baron et al., 1975). One-half of an applied dose was excreted from the bodies of rats within 3–4 days (Seidler et al., 1975).

After oral administration of gamma-HCH as a single dose of 60 mg/kg to rats, a maximal concentration of 8.8 ± 1.1 mg/kg tissue was reached in the brain after 12–24 h. This concentration decreased with a half-time of 1.5 days (Vohland et al., 1981). Oral administration of lindane in the diet for 13 weeks at doses up to 100 mg/kg diet resulted in concentrations in fat that were lower than those administered in the diet. The difference was more pronounced at higher doses. After administration of 100 mg/kg diet, 11.4 mg/kg were found in fat (Suter et al., 1983).

All studies in which lindane was fed continuously to rats showed that this compound does not accumulate in significant amounts in the body. The highest accumulation factor found for fatty tissues was about 2, and an average accumulation factor for fatty tissues of about 1 can be deduced from the published data. The corresponding factors for other tissues are considerably lower.

6.6 Biotransformation

6.6.1 Plants

In order to study the metabolism of lindane in wheat, plants were grown from seed containing 480 mg/kg of ^{14}C-lindane. In seedlings, 35.5% of the radiolabel was associated with unmetabolized lindane, 29.1% with the group of chlorobenzenes and 26.3% with chlorophenols. In mature plants, the extractable residues consisted of 5.4% lindane in roots and 21.4% in straw, up to 13.9% chlorobenzenes and up to 53% chlorophenols. The chlorobenzenes extracted from wheat roots were mostly tri- and tetrachlorobenzenes. The concentrations of di- and pentachlorobenzene and of gamma-PCCH were low (Balba & Saha, 1974).

Lindane and possible metabolites were determined in white cabbage after leaf application and in carrots grown in treated soil. In cabbage, a maximum of 0.04 mg/kg of lindane residues could be detected at the time of harvest. In the carrots, residue levels of 0.4–1.0 mg/kg were found. In the second year after treatment, residue levels were < 0.005 mg/kg. Up to 0.05 mg/kg of gamma-PCCH and traces of 1,2,4-tri- and 1,2,3,4-tetrachlorobenzene were found. Hexachlorobenzene was not detected in any of the samples (Eichler, 1975, 1980).

Itokawa et al. (1970) investigated the fate of ^{14}C-lindane in spinach and carrots grown in treated soil: 30–70% of the total residues in the different plant parts were lindane. Five metabolites were identified but not further characterized.

Lindane was converted to 36% soluble and 30% unextractable residues under outdoor conditions after foliar application to endives and lettuce.

About 97% of the soluble fraction was found to consist of chlorophenols in free or conjugated form. Minor quantities of various chlorobenzenes were found. The conversion to unextractable residues was dependent on weather conditions; the composition of the unextractable residues was not analysed in detail (Kohli et al., 1976a).

After lettuce plants were grown in nutrient solution containing ^{14}C-lindane at 1.45 mg/kg for 4 weeks, the radiolabel extracted from the plants consisted of about 77% lindane and 20% polar and 3% non-polar residues. 2,3,4,6-Tetrachlorophenol, conjugated tetra- and pentachlorophenol, and unidentified metabolites were found in the polar fraction. The non-polar fraction contained tri- and pentachlorobenzene as well as gamma-PCCH and HCCH (Kohli et al., 1976b).

The metabolism of lindane was investigated in a variety of plant-cell tissue cultures with high metabolic activity and in lettuce plants grown in nutrient solution (Stöckigt, 1976; Stöckigt & Ries, 1976). Tobacco tissue cultures were found to produce trace amounts of 1,2,4-trichlorobenzene, while in carrot cultures, 1,2,3,4-tetrachlorobenzene and several isomers of TCP conjugated with beta-glucose were found. This investigation also demonstrated that intact lettuce plants cannot produce pentachlorophenol or chlorobenzenes. Metabolism of lindane to carbon dioxide was not detected.

Moza et al. (1974) applied gamma-PCCH, a plausible metabolic intermediate of lindane, to young maize and pea plants in nutrient solution. Various chlorobenzenes and chlorophenols were formed. The most abundant metabolites in maize were 1,2,4,5-tetrachlorobenzene and 2,3,5- and 2,4,5-TCP, and the most abundant in pea plants was 1,2,4,5-tetrachlorobenzene.

Pea plants were grown under laboratory conditions in nutrient solution containing ^{14}C-lindane. After transfer into lindane-free medium, either lindane or its metabolites were released from roots, and to a lesser extent from other parts of the plant, within one day (Charnetski & Lichtenstein, 1973).

The results described above are reflected in Figure 3, which does not, however, include the conversion of lindane to traces of hexachlorobenzene or pentachlorophenol or to the corresponding alpha- and beta-HCH isomers (Kohli et al., 1976a,b; Steinwandter, 1976).

Kinetics and metabolism

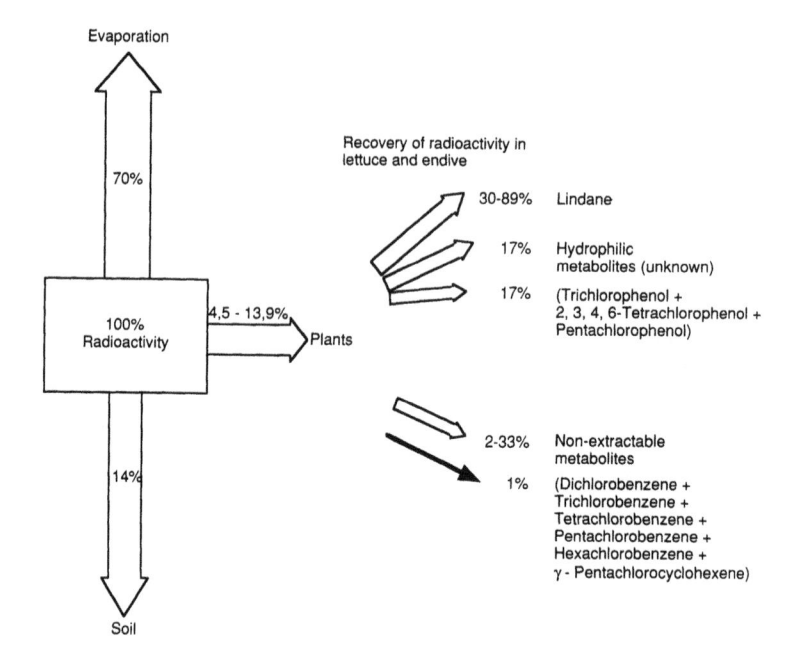

Fig. 3. Metabolism of lindane in lettuce and endive
From: Kohli et al. (1976a); Eichler (1980)

Appraisal

Lindane is not absorbed by the leaves of plants, and its poor absorption by roots rapidly reaches a plateau. Most of the lindane applied to plants is removed by evaporation or leaching. The rate of metabolic transformation is low. The main degradation pathway proceeds *via* formation of gamma-PCCH to TCP and tetrachlorophenol in free or conjugated form. Other metabolites that have been described occasionally, such as HCCH and chlorobenzenes, are present in only negligible quantities.

6.6.2 Microorganisms

The metabolism of lindane has also been investigated in bacteria, fungi, and algae. Chlorocycloalkenes, chlorobenzenes, and chlorophenols were found to be metabolic intermediates, and carbon dioxide to be the end-product. Volatile, chlorine-free hydrocarbons were also found (Haider & Jagnow, 1975; Haider et al., 1975).

Mixed populations of bacteria metabolize lindane to gamma-PCCH, alpha-, beta-, or gamma-3,4,5,6-tetrachloro-1-cyclohexene (TCCH), pentachlorobenzene, 1,2,3,4-, 1,2,3,5-, or 1,2,4,5-tetrachlorobenzene, 1,2,4- or 1,3,5-trichlorobenzene, 1,2- and 1,4- dichlorobenzene, as well as carbon dioxide (Yule et al., 1967; Haider et al., 1974; Kohnen et al., 1975; Mathur & Saha, 1975; Tu, 1975; Haider et al., 1976; Jagnow et al., 1977; Mathur & Saha, 1977; Vonk & Quirijns, 1979).

Metabolites of lindane were identified as PCCH and TCCH in populations of *Escherichia coli* (Francis et al., 1975; Vonk & Quirijns, 1979); PCCH, TCCH, 1,2,3,4-tetrachlorobenzene, and carbon dioxide in *Pseudomonas* sp. (Benezet & Matsumura, 1973; Matsumura et al., 1976; Engst et al., 1979a); and TCCH, 1,2,4-trichlorobenzene, and 1,4-dichlorobenzene in *Clostridium* sp. (Heritage & MacRae, 1977a,b; Ohisa & Yamaguchi, 1978b; Heritage & MacRae, 1979; Ohisa et al., 1980, 1982).

Quantitative data on the metabolism of lindane in bacteria are given by MacRae et al. (1967), Benezet & Matsumura (1973), Haider et al. (1974), Haider & Jagnow (1975), Kohnen et al. (1975), Mathur & Saha (1975, 1977), and Haider (1979). The most abundant metabolites are PCCH and TCCH (up to 45.8% and 21.7%, respectively, of the initial dose of lindane). Chlorobenzenes may occur in only small or trace amounts. Carbon dioxide is formed under aerobic or submerged incubation conditions, and up to 20% of an initial dose of lindane was converted to carbon dioxide within 140 days (Kohnen et al., 1975). Strictly anaerobic conditions resulted in rapid release of chloride from lindane and in its conversion to volatile chlorine- free metabolites. Within 5 days, up to 90% of the applied dose had been released as volatile, chlorine-free hydrocarbons (Haider & Jagnow, 1975).

Lindane was shown to be effectively metabolized in the algae *Chlorella* and *Chlamydomonas* (Sweeney, 1969; Elsner et al., 1972); 1,3,4,5,6-PCCH was reported to occur as a metabolite, but no quantitative data are available.

Kinetics and metabolism

Depending on the availability of oxygen, lindane may follow various metabolic pathways in bacteria (Fig. 4).

```
                    gamma-hexachlorocyclohexane
                            (C₆H₆Cl₆)

    anaerobic conditions              aerobic conditions

    dechlorination                    dehydrochlorination
       (-Cl₂)                              (-HCl)
         ↓                                   ↓
gamma-tetrachlorocyclohexene      gamma-hexachlorocyclohexene
        (C₆H₆Cl₄)                          (C₆H₆Cl₅)

                    dechlorination, dehydrochlorination
                            (-Cl₂, -HCl)
                                ↓
                          chlorobenzenes
              (Cl₄-, Cl₃-, Cl₂-, and Cl-benzene)
                                ↓
                          aerobic conditions
                               (+O₂)
                          carbon dioxide
                              (CO₂)
```

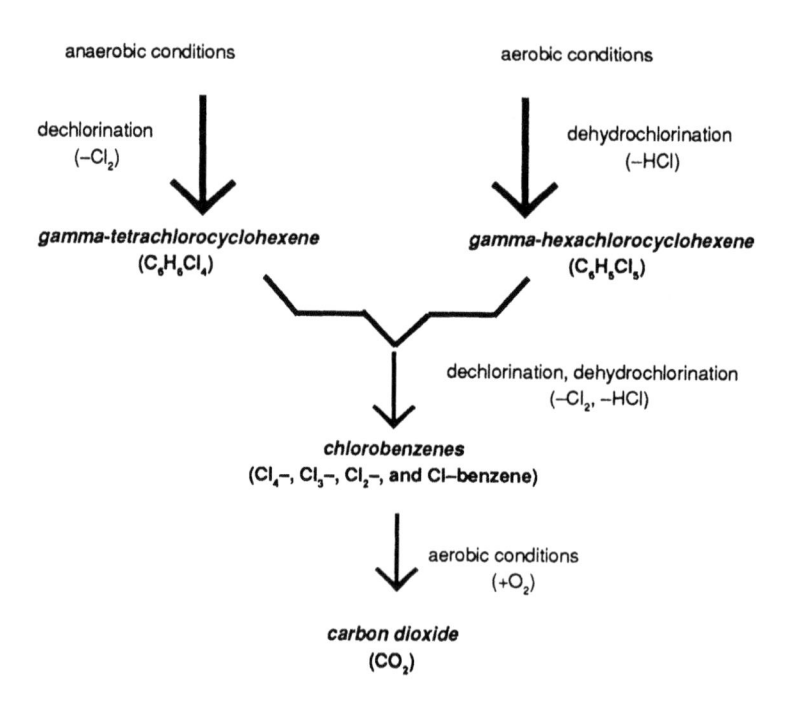

Fig. 4. Schematic metabolic pathway of lindane in bacteria

Unspecified fungi were also able to metabolize lindane, although at a lower rate than bacteria. The following metabolites were identified: gamma-PCCH; hexachlorobenzene; pentachlorobenzene; TCCH; 1,2,3,4-, 1,2,3,5-, and 1,2,4,5-tetrachlorobenzene; 1,2,3-, 1,2,4-, and 1,3,5-trichlorobenzene; 1,2- and 1,4-dichlorobenzene; pentachlorophenol; 2,3,4,5-, 2,3,4,6-, and 2,3,5,6-tetrachlorophenol; 2,3,4- and 2,4,6-TCP, and carbon dioxide. Metabolic intermediates such as PCCH were found at up to 1% of the initial dose of lindane. About 1% of the intial dose was converted to carbon dioxide after an incubation period of 52 days (Engst et al., 1974; Kujawa et al., 1976; Engst et al., 1977).

6.6.2.1 Anaerobic conditions

The influence of growth conditions on the metabolic route of lindane in bacteria was demonstrated in the facultative anaerobe *E. coli* as well as with mixed populations of soil microorganisms (Mathur & Saha, 1977; Vonk & Quirijns, 1979). These reports and others demonstrate the predominant formation of gamma-TCCH under anaerobic growth conditions. Anaerobic metabolism consists of a series of dechlorinating steps, leading to rapid formation of chlorine-free, volatile hydrocarbons and chloride (Haider & Jagnow, 1975; Jagnow et al., 1977). Carbon dioxide is not formed under anaerobic conditions.

A possible metabolic pathway under anaerobic conditions was proposed by Ohisa et al. (1980). Lindane is dechlorinated by a cytochrome P-450-dependent reaction to TCCH, followed by dechlorination to the unstable dichlorocyclohexadiene and dehydrochlorination to monochlorobenzene. The degradation of lindane serves as an energy source for the cells (Ohisa et al., 1982). A relationship between the metabolization of lindane and the Stickland reaction (a coupled oxidation–reduction reaction between pairs of amino acids) has been discussed (Ohisa & Yamaguchi, 1979; Ohisa et al., 1980, 1982). The ability to degrade lindane is linked to a bacterial enzyme system that catalyses the evolution of hydrogen during fermentation (Jagnow et al., 1977).

6.6.2.2 Aerobic conditions

Under aerobic conditions, the metabolism of lindane in bacteria is initiated predominantly by dehydrochlorination to gamma-PCCH (Vonk & Quirijns, 1979). Further intermediates are chlorobenzenes, and the end-product is carbon dioxide. No phenolic intermediate was observed under submerged conditions (Mathur & Saha, 1975).

6.7 Isomerization

The detection of beta-HCH in the tissues of rats fed gamma-HCH led to the conclusion that isomerization of lindane had occurred (Kamada, 1971); however, the purity of the gamma-HCH used was not given in this report, and it is possible that impurities were measured.

Studies in rats by Copeland & Chadwick (1979) and Eichler et al. (1983) demonstrated that lindane did not undergo bioisomerization. Gopalaswamy & Aiyar (1984) reported biotransformation of gamma-HCH to hexachlorobenzene in male rats. The results of a study by Chadwick & Copeland (1985), however, using six young female Fischer 344 rats administered lindane in arachis oil at 20 mg/kg body weight daily for 6 days (control animals received the vehicle only), indicated that no significant biotransformation of lindane to hexachlorobenzene occurred in these animals. The gamma-HCH contents of adipose tissue on a fat basis were 0.04 ± 0.003 mg/kg in controls and 129 ± 6 mg/kg in lindane-treated rats.

The possibility of isomerization of lindane to alpha- and beta-HCH was also investigated in mixed populations of soil microorganisms and other defined bacterial strains.

Newland et al. (1969) studied the degradation of lindane in simulated lake impoundments and found traces of alpha-HCH under anaerobic incubation conditions. Benezet & Matsumura (1973) described the formation of small amounts of alpha-HCH from lindane incubated either with aquatic sediments or with suspension cultures of *Pseudomonas putida* supplemented with NAD. Matsumura et al. (1976) described an NAD-dependent pathway in *P. putida* leading to the formation of alpha-HCH under anaerobic conditions. Three percent of the initial radioactivity of ^{14}C-lindane was found at a location with the same Rf value as alpha-HCH after separation of the metabolites by thin-layer chromatography.

Engst et al. (1979b) indicated isomerization of lindane to alpha- and beta-HCH in anaerobically grown cultures of *P. aeruginosa*, in a study in which metabolites were analysed by gas chromatography. Lindane was not reported to be isomerized to alpha- or beta-HCH in fungi (Engst et al., 1977).

Vonk & Quirijns (1979) found a conversion rate of lindane to alpha-HCH of 0.2% after anaerobic incubation of lindane for 4 or 8 weeks with either sandy or silt loam soil samples and *E. coli*. No alpha-HCH was formed in control experiments in which sterile nutrient medium was incubated with gamma-HCH for 28 days. Growing mycelium of *Aspergillus niger* produced no alpha-HCH. In this study, metabolites were identified by gas-liquid chromatography and verified by mass spectrometry.

Haider (1979, 1983) tested anaerobic, semi-anaerobic, and aerobic incubation of radioactive labelled lindane with *Citrobacter, Serratia, Clostridium, Klebsiella, Pseudomonas,* and *E. coli*. Incubation did not increase the level of alpha-HCH. The results of the experiment with *Pseudomonas* under anaerobic conditions were not reported.

Deo et al. (1981) studied the interconversion of gamma-HCH in a sterile aquatic solution over periods of 1 day up to 4 weeks. Gas-liquid chromatography indicated a slow interconversion of gamma-HCH with time. Solvent extracts were tested for their toxicity by topical application onto 2-day-old *Drosophila melanogaster*. The observed decrease in toxicity of the gamma-HCH solution with time may have been due to both degradation and isomerization to less toxic isomers, such as alpha-HCH.

Taken together, the results of tests conducted under anaerobic conditions show that only a very small amount of lindane, if any, is converted to alpha-HCH, and there is no conclusive indication of isomerization to beta-HCH.

7. EFFECTS ON LABORATORY MAMMALS AND IN *IN-VITRO* TEST SYSTEMS

Lindane has been tested for acute and for short- and long-term toxicity in a number of animal species. Some of the earlier studies were undertaken using material of unspecified purity; and in some others, technical-grade HCH was used that contained various quantities of alpha- and beta-HCH, in addition to lindane. Those studies that are relevant to this review have been included.

7.1 Single exposure

The acute toxicity of lindane has been investigated in numerous studies in a variety of species and strains of laboratory animals via several routes of application. The reported LD_{50} values for lindane given by different routes of administration are of the same order of magnitude in the various species, and no sex-dependent difference was seen. A marked difference in acute oral toxicity results from the type of vehicle used: oily solutions of lindane were more toxic than suspensions in water, and when mineral oils were used as carriers fewer toxic effects were seen than with vegetable oils (Muralidhara et al., 1979). Young animals were generally more sensitive than adults. Lindane was more toxic to animals suffering from protein deficiency than to rats with a normal protein supply (Chen, 1968).

7.1.1 Oral

The LD_{50} values for the mouse, rat, guinea-pig, and rabbit are summarized in Table 8.

The choice of vehicle used for administering lindane in studies of its acute oral toxicity is important: the LD_{50} after administration in an oily solution or of an emulsifiable concentrate was 88 mg/kg body weight, but that for wettable powders, granules, flowable concentrations and aqueous suspensions was in the order of 170 mg/kg or even higher.

Single oral doses of 40 mg/kg body weight dissolved in oil were lethal to dogs; dogs that received 30 mg/kg survived but had convulsions (Barke, 1950). McNamara and Krop (1948) found that a dose of 100 mg/kg body weight was lethal to all of three treated dogs, and 50 mg/kg caused death in four out of seven animals. These data indicate that the lethal dose for dogs of lindane administered as an oily solution is about 40–50 mg/kg body weight.

Table 8. Reported oral LD$_{50}$ values for gamma-HCH in experimental animals

Species	LD$_{50}$ (mg/kg)	References
Rats (male, female, or males and females)	90–270	Slade (1945); Woodard & Hagan (1947); Riemschneider (1949); Antonovic (1958); Gaines (1960); Edson et al. (1966); Muacevic (1966, 1970, 1971a,b); Chen (1968); Schafer (1972); Frohberg et al. (1972b)
Mice (different strains)	55–250	Woodard & Hagan (1947); Graeve & Herrnring (1951); Nurmatov (1965); Frohberg et al. (1972a); Paul et al. (1980); Wolfe & Ralph (1980)
Guinea-pig	100	Cameron (1945)
Rabbit	90–200	Cameron (1945); Nurmatov (1965)

In an incident in which eight cows ingested a powder containing 19.1% gamma-HCH, those that ate 112 g or more of the powder died, while those given 70 g survived. These findings indicate that, for cows, the fatal dose was between 70 and 112 g or 140–225 mg/kg body weight of the powder (equivalent to 28–45 mg/kg body weight) (McParland et al., 1973).

7.1.2 Intraperitoneal and intramuscular

Mice of the NMRI-EMD strain (SPF) were administered lindane as a 0.5% suspension in 0.5% carboxymethylcellulose solution and were observed for 14 days. The intraperitoneal and intramuscular LD$_{50}$ values were found to be 97 and 152 mg/kg body weight, respectively (Frohberg et al., 1972a). In rats, only the intraperitoneal LD$_{50}$ has been determined: it was found to be 69 mg/kg body weight in Wistar-AF/HAN-EMD administered the compound in carboxymethylcellulose (Frohberg et al., 1972b).

7.1.3 Inhalation

Wistar (HAN/Boe) rats were exposed by whole-body exposure to lindane at (analytical) concentrations of 0, 273, or 603 mg/m^3 for 4 h. The average particle size was 0.4 µm, and the animals were observed for 14 days. Neither deaths nor abnormalities were found (Oldiges et al., 1980).

Effects on laboratory mammals and in in-vitro test systems

The 4-h acute LC_{50} for a lindane (99.6%) aerosol was determined by exposing four groups of five males and five female KFM-HAN Wistar rats by inhalation to aerosols containing lindane at 0.1, 0.38, 0.64, or 2.1 mg/litre; 50% or more of the particles had a diameter of less than 7 μm. The observation time was 22 days. At toxic doses, signs of neurotoxicity (curved body posture, paddling movements and spasms) were observed. The acute 4-h LC_{50} was found to be about 1600 mg/m³ for animals of each sex (Ullman et al., 1986d).

7.1.4 Dermal

The acute dermal toxicity for rabbits was 200–300 mg/kg (Medvedev, 1974; see Izmerov, 1983).

Sherman strain rats were given one dermal application of lindane (99%) dissolved in xylene, and no attempt was made to remove the compound during the observation time of 14 days. The LD_{50} was 1000 mg/kg body weight for males and 900 mg/kg body weight for females (Gaines, 1960).

Male New Zealand rabbits, both young adult (2–3 kg) and just weaned (1 kg) were shaved only, shaved and depilated or shaved, depilated and 'stripped', and a commercial preparation of 1% lindane and 99% inert material was applied once to the entire body except the head, limbs, and perineal surface at a dose of 6 ml/kg body weight (equivalent to a dose of lindane of 60 mg/kg body weight, a dose reportedly used for scabies control in infants). The lindane was allowed to remain on the skin during the experiment. Two of four adult rabbits that were treated after having been shaved, depilated, and 'stripped' exhibited excitement after about 24 h. Adult rabbits that had been shaved only showed no effect. Weanling rabbits exhibited severe anorexia and convulsions, and death occurred in some cases. The effects were more pronounced in weanlings with inflamed or damaged skin. The concentrations of lindane in whole blood of weanlings when convulsions occurred (about 24 h after treatment) were 0.7–2.5 μg/ml (Hanig et al., 1976).

7.2 Short-term exposure

7.2.1 Oral

7.2.1.1 Mouse

In young dd mice fed diets containing lindane at 0, 2, 4, or 10 mg/kg diet for three months, no effect on growth and no histopathological change in the main organs were seen at any dose level (Chen & Liang, 1956). Similarly, Kitamura et al. (1970) saw no difference in behaviour, food consumption or body weight gain from that in controls in ICR mice fed diets containing lindane at 0.1, 1, 10, and 100 mg/kg diet for 36 days. No histopathological examination was carried out.

7.2.1.2 Rat

Short-term studies carried out by Slade (1945) and Laug (1948) were more or less inadequate for an evaluation.

Doisy & Bocklage (1949, 1950) fed lindane-containing diets to weanling rats for four weeks; doses of 400, 600, and 800 mg/kg diet caused high mortality rates. Food intake and weight gain were markedly reduced, especially in the group receiving 800 mg/kg diet. The animals showed irritability, hyperactivity, and convulsions. A dose of 200 mg/kg diet was without effect. Young rats were more susceptible than adults.

In a three-month toxicity study, groups of 15 male and 15 female Wistar KFM-Han (outbred) SPF rats were fed diets containing lindane (99.85%) at 0, 0.2, 0.8, 4, 20, or 100 mg/kg diet. After 12 weeks of treatment, most of the animals were sacrificed; the remaining rats were placed on a control diet for six weeks and then sacrificed. Lindane had no effect on mortality, food consumption, haematological parameters, the results of urinalysis, or clinical symptoms, although rats fed 100 mg/kg diet gained 8.4–14.9% less weight than controls. Liver cytochrome P-450 levels were increased in females given diets containing lindane at 0.8 mg/kg diet or more and in high-dose males at the termination of dosing; these values returned to the control levels during the recovery period. Such increases in cytochrome P-450 activity are regarded as an adaptation phenomenon due to induction of the microsomal detoxifying enzymes. Slight, dose-related, reversible increases in absolute and/or relative weights of livers and kidneys were observed in male and female rats fed lindane at 20 or 100 mg/kg diet, and histopathological examination revealed changes in these animals.

Effects on laboratory mammals and in in-vitro test systems

Those in the liver included dose-dependent, minimal-to-slight centrilobular hepatocellular hypertrophy at the end of the application period. After the recovery period, liver weights were found to be normal, and no centrilobular hypertrophy was seen. In the kidneys, minimal-to-slight, unicellular and multicellular necrosis of epithelial cells was observed in proximal convoluted tubules, and basophilic tubules, interstitial nephritis and hyaline droplets were seen in epithelial cells of the convoluted tubules. After the recovery period, the tubular degeneration was no longer present, but the nephritis and basophilic tubules were still present in the animals that had received 100 mg/kg. No effect was observed with doses of 4 mg/kg diet (equivalent to 0.3 mg/kg body weight) and below (Suter et al., 1983).

In a 12-week study with groups of 10 male and 10 female Wistar RIV:TOX (C-S) rats, four weeks old at the beginning of the experiment, lindane (99.8%) was administered in the diet at concentrations of 0, 2, 10, 50, and 250 mg/kg. At the highest dose, increases were seen in the induction of enzymes, such as aminopyrine-N-demethylase and ethoxyresorufin-O-deethylase, but cytochrome P-450 and aryl hydroxylase activity were not increased. At the two highest dose levels, the weights of livers, kidneys, and thyroid were increased. The NOEL of lindane in this study was 10 mg/kg diet (equivalent to 0.75 mg/kg body weight) (van Velsen et al., 1984).

Young male Wistar rats were fed gamma-HCH at a dose of 0 (five rats) or 800 mg/kg diet (eight rats) for two weeks, and urinary excretion of body constituents that reflect renal function was measured. Glucosuria and increased excretion of creatinine and urea were found, and hypertrophy and degeneration of the renal tubular epithelia were observed histologically (Srinivasan et al., 1984). In young male Wistar rats administered gamma-HCH at 800 mg/kg diet for two weeks, liver weights were increased, but no difference was found in moisture, nitrogen, protein, or glycogen levels. The fat and DNA content of the liver were found to be increased, but the DNA content per unit issue was decreased. The predominant change in the liver was hypertrophy. Testicular weight was no different from that in control animals, but the protein content was higher, and the DNA content was lower. The histological changes observed were tubular atrophy and spermatogenic arrest; the interstitial space was found to be oedematous (Srinivasan et al., 1988).

Liver function was studied in male Wistar rats fed a control diet (6–8 rats) or gamma-HCH at 800 mg/kg diet (8-12 rats). Gamma-HCH produced noticeable hepatocellular effects, as indicated by increased activity of serum aminotransferases, hepatic glucose-6-phosphate dehydrogenase and aldolase and decreased activity of liver glucose-6-phosphatase. Liver mitochondrial dinitrophenol/Mg^{++}/Ca^{++}-activated ATPase activity was decreased, and levels of microsomal Na^+, K^+-ATPases were lower in treated than control animals (Srinivasan & Radhakrishnamurty, 1988).

In a preliminary study, 24 Fischer 344 weanling, female rats, received daily oral doses of either arachis oil or lindane at 0.069 mmol/kg body weight for 189 days. Lindane induced a significant increase in body weight after 112 days of treatment. In a subsequent dose-response study, female Fischer 344 rats, 21 days of age, were gavaged daily with arachis oil (six rats) or lindane at 5 (six rats), 10 (eight rats), 20 (12 rats), or 40 mg/kg body weight (12 rats). At 20 mg/kg, lindane induced an increase in body weight after 10 weeks of treatment. At 40 mg/kg, 7 out of the 12 rats died; the other animals had increased body weight gain. Greater food consumption was observed, and obesity was induced, as indicated by the Lee index. In addition, lindane caused delay in vaginal opening, disrupted oestrous cycling, reduced pituitary and uterine weights and elevated food consumption during pro-oestrous. This response suggests that, by inducing alterations in the reproductive function of female rats and by interfering with hormonal regulation of the energy balance, lindane may be anti-oestrogenic rather than oestrogenic as previously proposed (Chadwick et al., 1988).

7.2.1.3 Dog

Lehman (1952) found a high mortality rate in dogs given lindane at daily doses of 10 or 15 mg/kg body weight on 5 days per week over a period of 2–221 days. (Details not available.) Lehman (1965) reported a study initially conducted by Fitzhugh et al., in which dogs (two males and two females per group) were exposed to lindane at 0 or 15 mg/kg diet (equivalent to approximately 0.6 mg/kg body weight) in the diet for a total of 63 weeks. No effect was observed on mortality, organ weights, body weight gain, haematological parameters, or histological appearance.

During a two-year toxicity study (Rivett et al., 1978), groups of four male and four female beagle dogs were fed lindane (99%) at 0, 25, 50, or 100 mg/kg diet. The amounts of lindane that were actually ingested were 0.83, 1.60, and 2.92 mg/kg body weight per day. Convulsions seen occasionally in control and low-dose animals were not related to the treatment. No treatment-related change was observed in body weight, food or water consumption, ophthalmological parameters, electroencephalographic traces, results of haematological examinations, urinalysis, and liver function tests, or organ weights. At autopsy, somewhat darker colouration and a brittle consistency of the liver were seen at 100 mg/kg. In addition, alkaline phosphatase activity was increased in the highest dose group. No treatment-related abnormality was apparent with 25 or 50 mg/kg diet.

A supplementary group of four male and four female dogs was administered lindane at 200 mg/kg diet for 32 weeks. High-voltage slow-wave activity changes, possibly indicative of nonspecific neuronal irritation, were recorded in electroencephalographic tracings at this dose level. No such effect was observed in the two-year study at 100 mg/kg diet (Rivett et al., 1978).

7.2.1.4 Pig

Schnell (1965) fed diets containing lindane (99.5%) at 0, 5, 10, 20, 40, or 80 mg/kg diet to groups of five pigs over a period of nine months. No clinical symptom was seen in any animal during the test period. Food and water intake remained normal, and haematological investigations and histopathological examination of the liver, spleen, kidneys, adrenals, heart, and brain revealed no substance-related change, even at the highest dose level tested.

7.2.2 Inhalation

7.2.2.1 Mouse

Balaschow (1964) exposed mice for 6 h/day to a lindane aerosol containing a nominal concentration of 1 mg/m^3 for 2.5 months. During the first two weeks, white blood cell counts showed the presence of leuko-cytosis; from the end of the first month, leukopenia was observed, and toxic granulations and vacuoles appeared in the nuclei and cytoplasm of some leukocytes. Later, a reduced mitotic rate was observed. The relationship between the different cell types in the bone marrow was undisturbed.

Four groups of 45 male and 45 female CD-1 (Charles River) mice were exposed by whole-body inhalation to aerosols (geometric mean particle diameter, about 3 µm) containing lindane (purity at least 99.6%) at 0, 0.3, 1.0, or 5–10 g/m^3 for 6 h/day, on five days per week for 14 weeks. The test dose levels were selected on the basis of a preliminary range-finding study, which showed that males did not develop major signs of toxicity after five exposures for 6 h/day to lindane at 1.0 and 10.0 mg/m^3. During the main study, however, an unexpectedly high mortality rate was seen in females exposed to 10 mg/m^3; after the first five exposures, therefore, the concentration for the high-dose group was lowered to 5 mg/m^3. Subgroups of 15 mice of each sex in each group were sacrificed after 7, 14, and 20 weeks. The group sacrificed at 20 weeks was a recovery group, which was not exposed to lindane after exposure week 14. Ten of the 15 mice in each subset were used for pathological evaluation, and the remaining five were examined for serum lindane levels. The lindane aerosol was highly toxic to female mice at 5 mg/m^3 and also at 1 mg/m^3. The NOEL was concluded to be 0.3 mg/m^3 (Klone & Kintigh, 1988).

7.2.2.2 Rat

Groups of 12 male and 12 female Wistar Han/Boe SPF rats were exposed by whole-body inhalation to lindane (99.9%) at nominal concentrations of 0, 0.02, 0.12, 0.6, or 4.5 mg/m^3 (average particle size, 0.92 µm) for 6 h/day for three months. The two groups that received 0 or 4.54 mg/m^3 were used to investigate recovery. Slight diarrhoea and ruffled fur were observed temporarily in the high-dose animals only. Measurements of body and organ weights and of food and water intake, clinical chemistry, and histopathology showed no treatment-related change. Hepatic cytochrome P-450 values were increased at the end of the exposure period in animals at the highest dose, but all values returned to that of the control during the six-week recovery period. Increased kidney weights and cloudy swelling of the tubular epithelium were observed especially in males at the two highest dose levels, but, again, all values were comparable with those in controls at the end of the recovery period. The NOEL was probably 0.6 mg/m^3 (Oldiges et al., 1983).

7.2.3 Dermal

Four groups of 49 male and 49 female Charles River rats (Crt: (WI)BR strain) were exposed dermally to lindane (purity at least 99.5%) at dose levels of 0, 10, 60, or 400 mg/kg per day, selected on the basis of a preliminary range-finding study, on five consecutive days per week. The test substance was applied as a suspension in aqueous carboxymethyl cellulose at a constant volume of 4 ml/kg to a clipped area of the skin on the dorsal area and was retained for 6 h with a dressing consisting of a gauze pad heat-welded to plastic-backed aluminium foil. In the main phase of the study, 23 animals of each sex in each group were treated for 13 weeks before sacrifice; one group of 13 animals/sex were sacrificed after 6 weeks of treatment; a third (recovery phase) group, consisting of 13 animals of each sex, was retained in the study for an additional six weeks. At sacrifice, three animals in each group were selected for determination of tissue levels of lindane (Brown, 1988).

The toxic effects induced by subchronic dermal exposure to 60 and 400 mg/kg consisted of pathological lesions of the kidneys in males (increased organ weight, hyaline droplet formation, tubular degeneration with necrosis, basophilic tubules, casts) and hypertrophy of the liver in males and females. Whereas the effects on the liver were reversible, some of the histopathological changes in the kidneys persisted (tubular degeneration with necrosis, granular casts) after the recovery period. Although there was evidence of increased intensity of hyaline droplet formation at the lowest dose tested (10 mg/kg per day), this effect was very slight; that level could therefore be considered to be the NOEL. The use in this study of semi-quantitative methods for determination of blood levels, protein, and turbidity makes it difficult to come to any definite conclusion; however, the results of the urinalysis did not provide evidence that lindane adversely affects kidney function.

7.3   Skin and eye irritation; sensitization

7.3.1   *Primary skin irritation*

Application of 0.5 g of lindane to the intact skin of New Zealand white rabbits, in a study performed in compliance with the guidelines of the Organization for Economic Co-operation and Development and the US Environmental Protection Agency, did not cause irritation (Ullmann et al., 1986a).

7.3.2   *Primary eye irritation*

Lindane placed in the conjunctival sac of the left eye of New Zealand white rabbits at 0.1 g was slightly irritating (Ullmann et al., 1986b).

7.3.3   *Sensitization*

The allergic potential of lindane was tested in a Magnusson-Kligman maximization test (according to the guidelines of the Organization for Economic Co-operation and Development) on Dunkin-Hartley albino guinea-pigs. Ten males and ten females received lindane (99.6%) and five males and five females received the vehicle, ethanol. No difference was seen between the test group and the controls after the first and second challenge applications 24 and 48 h later, and it was concluded that lindane has no skin sensitizing (contact allergenic) potential in these guinea-pigs (Ullmann et al., 1986b).

Ullmann et al. (1987a) conducted a further maximization test (following the guidelines of the Organization for Economic Co-operation and Development) with Dunkin-Hartley albino guinea-pigs to test the contact hypersensitization potential of a lindane formulation. Ten males and ten females received intradermal injections of 5% 'Nexit fluessig' (containing 25.9% lindane) in saline, and five males and five females received the saline vehicle. No sensitization reaction was observed after the first and second challenge applications, 24 and 48 h later.

Comparable experiments were carried out with two other formulations, 'Nexit stark', a powder containing 78.9% lindane (Ullmann et al., 1987b), and 'Agronex Saatgutpuder', a powder containing 20.1% (Ullmann et al., 1987c). Each was administered as intradermal injections of 0.1% in saline. No sensitization reaction was observed after two challenge reactions 24 and 48 h later.

7.4 Long-term exposure

7.4.1 Oral

In two long-term studies, lindane powder was mixed into the diet of Wistar rats (10 males and 10 females) at 10, 100, or 800 mg/kg diet and as an oily solution at 5, 10, 50, 100, 400, 800, or 1600 mg/kg diet, either as the gamma isomer or as technical HCH (containing only 13% of the gamma isomer). Two control groups were used. At 100 mg/kg diet, liver weight was increased, and histopathological examination revealed hepatocellular hypertrophy, fatty degeneration and necrosis as well as nephritic reactions (granular degeneration and calcification in male rats). These findings were more pronounced at the 400, 800, and 1600 mg/kg dietary levels. At these concentrations, the life span of animals in groups treated with the oily solution was shortened by 20–40%. The NOEL in this experiment was 50 mg/kg diet (Fitzhugh et al., 1950; Lehman, 1952).

Similar results were obtained in another lifetime study, in which groups of 10 male and 10 female rats received lindane at 25, 50, or 100 mg/kg diet. The dose of 25 mg/kg had no effect on the liver, but hepatocellular hypertrophy was observed with 50 mg/kg and slight fatty liver-cell degeneration was described in the group receiving 100 mg/kg diet (Truhaut, 1954).

7.4.2 Appraisal of acute and short- and long-term studies

The acute oral toxicity (LD_{50}) of lindane in different species, depending on the vehicle used, ranges from 56 to 480 mg/kg body weight. Preparations in oil were more toxic than aqueous solutions or suspensions. The ranges for rats and mice were similar (88–270 and 56–246 mg/kg, respectively). The dermal LD_{50} for rats is approximately 900 mg/kg body weight, but smaller amounts (60 mg/kg as a 1% cream) caused convulsions, anorexia, and deaths in weanling rabbits. No skin irritation or sensitization was observed, and eye irritation was slight.

Although older, long-term studies in the rat suggest a NOAEL of 25 mg/kg diet, contemporary short-term studies in rats indicate that this level is 10 mg/kg diet, equivalent to 0.75 mg/kg body weight, on the basis of increased hepatic, renal, and thyroid weights, increased cytochrome P-450 activity and histopathological findings in liver and kidneys.

7.5 Reproduction, embryotoxicity, and teratogenicity

7.5.1 Reproduction

Trifonova et al. (1970) found no reduction in the fertilization rate of female rats after oral treatment for 90 days with lindane at approximately 5 mg/kg body weight. When the dose was doubled over a test period of 138 days, the fertilization rate was reduced. (Details not given.)

A three-generation test was carried out in which 10 male and 10 female CD-rats were administered lindane at concentrations of 25, 50, or 100 mg/kg diet continuously. The treatment had no influence on fertility, litter size, breeding rate, weight of newborn animals, lactation, malformation rate, or maturation. The liver weights of young animals of the F_3b generation were increased, especially among females. Histopathological examination of the liver showed enlarged hepatocytes and vacuolization in animals treated with 50 and 100 mg/kg diet (Palmer et al., 1978a).

7.5.2 Embryotoxicity and teratogenicity

7.5.2.1 Oral administration

Mouse: Lindane was administered orally to seven groups of 25 pregnant NMRI-EMD (SPF) female mice at 0, 12, 30, or 60 mg/kg body weight in 0.5% carboxymethyl cellulose, on either days 6–15 or days 11–12 of pregnancy. In the group receiving the highest dose, fetal mortality was increased and fetal weights were decreased. A slight, non-dose-related increase in malformation rate was found in the mid-dose group (4.2%, as compared to 1.9% in controls). At the highest dose, increased maternal mortality (48%) and reduced body weight gain were observed. The treatment had no effect on the number of implantations per dam, the percentages of early and late resorptions, the number of runts or the malformation rate (Frohberg & Bauer, 1972b).

Rat: Groups of 20 female CFY-rats received lindane at 5, 10, or 20 mg/kg body weight by gavage during days 6–15 of pregnancy. In the groups given 10 and 20 mg/kg, maternal toxicity (reduced food intake and reduced weight gain) was observed, and two female rats given 20 mg/kg died. At the same dose, there was a dose-related increase in the incidence of offspring with extra (14th) ribs, which was statistically significant. Other anomalies and litter parameters were comparable to those in the controls, and there was no evidence of embryo- or fetoxicity (Palmer et al., 1978b).

Khera et al. (1979) gave female Wistar rats (20 animals per group) a lindane formulation (50% in corn oil) at 3.12, 6.25, or 12.5 mg/kg body weight (expressed as 100% lindane) by intubation on days 6–15 of gestation. No effect was seen on the number of living fetuses per litter, the number of dead plus resorbed fetuses or mean fetal weight at 22 days of gestation. No malformation other than the usual range of developmental variants was observed in any group. A slight increase in the frequency of anomalies of the ribs and reduced cranial ossification were seen in the fetuses exposed to 6.25 mg/kg; these effects were confined to two litters and were probably not dose-related.

Female rats that received lindane orally at a dose of 25 mg/kg body weight daily during pregnancy had higher post-implantation embryonal mortality than controls, and at 12.5 mg/kg no mortality was found. Neither dose level induced teratological abnormalities (Mametkuliev, 1976; see Izmerov, 1983).

Rabbit: Lindane was administered by intragastric intubation to New Zealand white rabbits (13 animals per group) on days 6–18 of gestation at doses of 5, 10, and 20 mg/kg body weight. All treated animals showed slight tachypnoea and lethargy during the treatment period, and body weight gain and food intake were reduced. Pre-implantation loss was significantly higher in the group given 20 mg/kg, but, as treatment did not start until day 6 of gestation, this effect is unlikely to have been due to lindane. Post-implantation loss and the incidence of resorptions were increased at 5 and 20 mg/kg. The number of offspring with extra (13th) ribs was significantly lower in animals given 5 mg/kg and significantly higher in rabbits at 20 mg/kg than in controls. Fetal and litter weights were unaffected, and the incidence of other anomalies was similar to that in controls (Palmer et al., 1978b).

Dog: An increased frequency of stillbirths, unrelated to dosage or period of administration, was seen in beagle dogs fed lindane at 0 (five dogs), 7.5 (13 dogs) or 15 mg/kg body weight (14 dogs) from day 1 or 5 throughout gestation. No significant teratogenic effect was observed. The number of living pups was similar in control and test groups (Earl et al., 1973).

Pig: Groups of six female pigs received lindane at 0, 50, or 500 mg/kg diet from 30 days prior to mating until day 30 of gestation. No treatment-related effect was found on number of embryos, embryo weight, or rate of ovulation (Duee et al., 1975).

Cow: In an accidental poisoning incident, four pregnant cows received lindane at 13.4 g (28 mg/kg body weight) 6–17 weeks pre-partum. All had convulsions and muscular tremors in the ensuing 48 h but recovered with veterinary treatment. All calved on time and produced normal, healthy calves. Four non-pregnant cows died after receiving 21 g of lindane, which suggests that the minimum lethal dose is 28–45 mg/kg body weight (McParland et al., 1973).

7.5.2.2 Subcutaneous injection

Mouse: Lindane was administered subcutaneously (in a 0.5% carboxy-methyl cellulose solution) to groups of 25 pregnant NMRI mice at 6 mg/kg body weight on either days 11–13 or days 6–15 of pregnancy. Except for a slight increase in the frequency of runts in the latter group, no effect was seen on the number of implantations or of living embryos per dam or on the percentage of absorptions or resorptions; no treatment-related malformation was reported (Frohberg & Bauer, 1972a).

Rat: Groups of 20 Sprague-Dawley rats received lindane at doses of 0, 5, 15, or 30 mg/kg body weight by subcutaneous injection on days 6–15 of gestation. Maternal toxicity was observed in the mid- and high-dose groups. No effect attributable to the administration of lindane was noted on pregnancy rates, maternal gross pathology or reproduction, or offspring viability and development. No teratogenic effect was found at necropsy or in visceral and skeletal examinations (Reno, 1976a; Hazelton Laboratories, 1976a).

Rabbit: Lindane was injected subcutaneously at 0, 5, 15, 30 or 45 mg/kg body weight into pregnant rabbits on days 6–18 of pregnancy, except that the highest dose was given on days 6–9 and then reduced to 30 mg/kg body weight. No embryotoxic or teratogenic effect was found in fetuses exposed to the two lower doses. At the two higher dose levels, increased maternal toxicity was found. At the highest dose, the number of resorptions was increased, and 14 out of 15 animals died (Reno, 1976b; Hazelton Laboratories, 1976b).

7.5.3 Reproductive behaviour

Adult female Fischer (CDF-344) rats were injected intraperitoneally on the morning of pro-oestrus with lindane at 25, 33, 50, or 75 mg/kg body weight in sesame oil, and in the evening they were examined for lordosis behaviour with a sexually experienced male. A dose-dependent reduction in sexual receptivity was seen with increasing doses of lindane: treated animals required a greater number of mounts before the first lordosis response was observed, and they may have required more sensory stimulation to elicit the lordosis reflex. Most of the females also failed to exhibit proceptive behaviour (darting and hopping) during the mating test. This inhibition resembles the rapid effects of another chlorinated pesticide, chlordecone, and does not appear to depend upon disruption by lindane of the inhibition of the central nervous system by gamma-aminobutyric acid (GABA). The results substantiate previous suggestions that the ability of chlorinated pesticides to interfere with intracellular oestradiol receptors cannot explain their rapid attenuation of reproductive behaviour (Uphouse, 1987).

7.5.4 Appraisal of reproductive toxicology

Lindane was investigated in tests covering all aspects of reproduction (three-generation studies in rats) and in tests for embryotoxicity and teratogenicity by oral, subcutaneous and intraperitoneal administration in mice, rats, dogs, and pigs.

Lindane did not exhibit teratogenic properties after oral or parenteral application (extra ribs were regarded as variations). Fetal and/or maternal toxic effects were observed in rats with doses of 10 mg/kg body weight and higher given by oral gavage; 5 mg/kg is therefore considered to be the NOAEL.

No effect on reproduction or maturation was seen in the three-generation study at doses of lindane up to 100 mg/kg diet, but morphological signs suggesting liver enzyme induction occurred with doses from 50 mg/kg diet in the third generation. The NOEL in this test was 25 mg/kg diet (equivalent to approximately 1.25 mg/kg body weight).

7.6 Mutagenicity and related end-points

Lindane was tested in mutagenicity tests with a variety of end-points. The relevant experiments are summarized in Tables 9, 10, and 11. Those studies that were not performed according to protocols which comply to the present international standards are considered to be of limited relevance; some studies used lindane preparations of less than 99% purity or of unknown purity. The results are so consistent, however, that the limitations of some studies did not vitiate a final assessment.

7.6.1 DNA damage

The ability of lindane to damage DNA was tested in *Bacillus subtilis* and in *Escherichia coli* WP2 in the *rec* assay, and tests for unscheduled DNA synthesis tests were performed in primary rat hepatocytes and human fibroblasts. No mutagenic potential was detected.

Sina et al. (1983) developed a sensitive alkaline elution assay in non-radiolabelled rat hepatocytes to measure DNA single-strand breaks induced by chemicals. This assay is used to predict carcinogenic/mutagenic activity. Lindane at doses of 0.03 and 0.3 mmol/litre induced DNA damage, increasing with dose.

After oral administration of lindane to rats and mice, a very low covalent binding index (0.02–0.01) was calculated, suggesting that no significant binding to DNA had occurred.

The incorporation of orally administered radiolabelled thymidine into liver DNA was determined in SIV-50-SD rats 24 h after a single oral dose by gavage of 0.01, 0.1, or 1.0 mmol/kg gamma-HCH. No effect on liver DNA synthesis was seen (Büsser & Lutz, 1987).

7.6.2 Mutation

The ability of lindane to induce gene mutation has been investigated extensively in *S. typhimurium* and *E. coli*, using an adequate range of strains to cover both base-pair and frame-shift mutations. Most of the tests were performed both with and without metabolic activation by 9000 × *g* preparations from the livers of induced rats or mice (Table 9).

Negative results were obtained in the host-mediated assay using mice and *S. typhimurium* or *Serratia marescens*. Furthermore, a test for point mutations in V79 Chinese hamster cells, the *hprt* test for forward mutations, indicated no mutagenic effect of lindane. A test for sex-linked recessive lethal mutation in *Drosophila melanogaster* also gave a negative result (Table 9).

D. melanogaster were also used to test for dominant lethal mutation. Groups of 25 males and 25 females aged 6–24 h were transferred to food containing HCH (Gammexane) at 20 mg/kg food medium, and their progeny were raised on this food. Five males and five females of the F_1 generation (the 'toxic generation') were raised on normal food and were allowed to mate with each other and lay eggs for 24 h in 10 oviposition jars. From this generation of flies, three successive mutation-generations were raised on normal food, and the numbers of larvae hatched from eggs laid on each of the first 10 days after enclosion were again recorded. The percentage of larvae hatched from the total number of eggs laid, cumulated over the entire period, was significantly decreased in the second and third generations. These results suggest that the preparation tested is mutagenic (Sinha & Sinha, 1983).

A test for induction of reverse mutations in *Saccharomyces cerevisiae* gave inconclusive results.

7.6.3 Chromosomal effects

Most of the cytogenetic tests performed with lindane both *in vivo* and *in vitro* did not indicate mutagenic properties of lindane. In only one study were there positive findings, but the purity of the material tested was not given and the description of the test was poor. Lindane therefore apparently does not induce chromosomal breakage (Table 10).

Lindane was tested for its ability to induce sister chromatid exchange *in vivo* (in mice by oral and intraperitoneal administration) and *in vitro* (in Chinese hamster ovary cells); no effect was seen. No mutagenic effect was observed in a test for micronucleus formation in the bone marrow of mice treated *in vivo*, and lindane did not induce chromosomal damage *in vivo*.

Effects on laboratory mammals and in in-vitro test systems

Table 9. Result of mutagenicity tests of gamma-HCH

Test system (Organisms/Cells)	Dose	Type of test	Metabolic activation	Result	Reference
Bacillus subtilis					
H17 rec⁺	0.02 ml of solution containing 1 mg/ml in DMSO	plate	none	–	Shirasu et al. (1976)
M45 rec⁻		plate	none	–	
Escherichia coli					
WP2 try⁻	approx. 1 mg	plate	none	–	Ashwood-Smith et al. (1972)
WP2	4 gradient plates, covering 10 000-fold concentration range	plate	S9-mix	–	Probst et al. (1981)
WP2 uvr A⁻		plate	S9-mix	–	
WP2 urv A	6–7 dose levels up to 5000 µg	plate S9-mix	none –	–	Oesch (1980)
Salmonella typhimurium					
TA98, TA100, TA1535, TA1537, TA1538, TA1950, TA1978	1–1000 µg	plate	S9-mix	–	van Dijck & van de Voorde (1976)
TA98, TA100, TA1535, TA1538	93, 139, 208 µg	plate	none	–	Röhrborn (1977a)
TA100, TA1535, TA1537, TA98	6–7 dose levels up to 5000 µg	plate plate	none S9-mix	– –	Oesch (1980)
TA100, TA1535, TA1537, TA98	8 dose levels, 0 up to 333 µg	plate	none S9-mix	– –	Haworth et al. (1983)

Table 9 (contd)

Test system (Organisms/Cells)	Dose	Type of test	Metabolic activation	Result	Reference
Salmonella typhimurium (contd)					
TA100, TA1535, TA1538, TA98	4, 20, 100, 500, or 2500 µg in DMSO	plate	S9	–	Anderson & Styles (1978)
G46, TA98, TA100, TA1537, TA1538, C3076, TA1535, D3052	concentration gradient	plate	S9-mix	–	Probst et al. (1981)
Host-mediated assay					
Salmonella typhimurium G46	25 mg/kg bw (subcutaneous)	mouse (NMRI)	nr	–	Buselmaier et al. (1972)
Serratia marescens a 21. *leu*⁻	25 mg/kg bw (subcutaneous)	mouse (NMRI)	nr	–	Buselmaier et al. (1972)
Mammalian cells					
hprt locus V79 Chinese hamster cells	0.5–500 µg/ml 0.5–250 µg/ml	plate plate	S9-mix S9-mix	– –	Glatt & Oesch (1984); Oesch & Glatt (1984)
Sex-linked recessive lethal test					
Drosophila melanogaster	0.001% (aqueous sol.)	injected into abdomen (0.2 µl)	nr	–	Benes & Sram (1969)

DMSO, dimethyl sulfoxide; nr, not relevant; bw, body weight

Effects on laboratory mammals and in in-vitro test systems

Table 10. Results of tests for other genetic effects

End-point	Dose	Effects	Result	Reference
Chromosomal aberrations *in vitro*				
Chinese hamster fibroblast cell line (CHL)	2.1 mg/ml[a] (in ethanol)	Chromatid gaps, chromatid and chromosomal breaks	Equivocal	Ishidate & Odashima (1977)
Lymphocytes from human peripheral blood (different donors)	0.1, 0.5, 1.0, 5.0, or 10 µg/ml	Chromosomal breakage only at toxic dosages (5 and 10 µg/ml)	Equivocal[b]	Tzoneva-Maneva et al. (1971)
Chromosomal aberrations *in vivo*				
Chinese hamster bone-marrow cells	0.125, 1.25, or 12.5 mg/kg body weight orally for 5 days	Increase in chromosomal gaps at highest dose level	(−)	Röhrborn (1976, 1977a)
Syrian hamster bone-marrow cells	64, 128, 280, or 640 mg/kg body weight	No chromosomal aberration	−	Dzwonkowska & Hubner (1986)
Rat bone-marrow cells	1.5, 7.0, or 15 mg/kg body weight orally for 12 weeks	−	−	Gencik (1977)
Human lymphocytes	occupational exposure (no further details)		−	Desi (1972)
Sister chromatid exchange *in vivo*				
Mouse bone-marrow cells (strain CF1)	Male/female: 2/1.6, 10/8, or 50/40 mg/kg body weight as a single oral application	−	−	Guenard et al. (1984a)
Mouse bone-marrow cells (strain CF1)	Single intraperitoneal injection of 1.3, 6.4, or 32.1 mg/kg body weight	−	−	Guenard et al. (1984b)

Table 10 (contd)

End-point	Dose	Effects	Result	Reference
Micronucleus test *in vivo*				
Mouse erythroblasts (CBA male mice)	75 mg/kg body weight	–	–	Jenssen & Ramel (1980)
Dominant lethal test				
Rats (males; strain Chbb = THOM)	1.5, 7.0, or 15 mg/kg body weight daily for 8 weeks orally (in olive oil)[c]	–	–	Röhrborn (1977b)
Rat (males; strain Wistar)	1.5, 7.0, or 15 mg/kg body weight in olive oil	–	Questionable positive[b]	Cerey et al. (1975)
Mouse (males; strain ICR/Ha Swiss)	15, 75, 200, or 1000 mg/kg body weight once intraperitoneally	–	–	Epstein et al. (1972)
	15 mg/kg body weight five times, orally	–	Equivocal	Epstein et al. (1972)
Mouse (males; strain NMRI-EMD)	Single intraperitoneal injection of 12.5, 25, or 50 mg/kg body weight	–	–	Frohberg & Bauer (1972c)

[a] Maximal effective dose
[b] Inadequate study; protocol does not comply with international standards
[c] Males dosed continuously during the whole mating period (8 weeks)

Effects on laboratory mammals and in in-vitro test systems

Table 11. Results of tests for DNA damage

End-point	Dose	Type of test	Metabolic activation	Result	Reference
Unscheduled DNA synthesis (transformed SV-40) human fibroblast (cell-line VA-4)	1, 1000 µM (in acetone)	Tissue culture fluid	None	– S9-mix	Ahmed et al. (1977) –
Unscheduled DNA synthesis and repair capacity after damage by UV-rays (human lymphocytes)	500 µg/ml	Tissue culture	–	50–70% inhibition	Rocchi et al. (1980)
Primary rat hepatocytes	100 nmol/ml (in DMSO)	Plate	–	–	Probst et al. (1981)
Covalent DNA binding Male mouse (strain NMRI, CF1 and C6B3F1)	12–13 mg/kg body weight orally and 8.7 up to 23 mg/kg body weight orally	Liver DNA	Not relevant	– (covalent binding index, 0.02–0.1)	Sagelsdorff et al. (1983)

Two of three tests for induction of dominant lethal mutation in rats gave clearly negative results, and the other gave a questionably positive response. The significance of the latter test must be regarded as low, because the purity of the material tested was not given and the test was not performed in compliance with an acceptable standard.

7.6.4 Miscellaneous tests

Lindane tested in a MO_4 cell culture at doses of 1, 10, and 100 µg/ml induced no multinucleation or major toxicity (de Brabander et al., 1976).

Lindane at a concentration of 101.8 µg/ml did not induce 6-thioguanine-resistant mutations in Chinese hamster V79 cells. Concentrations of 100 and 200 µg/ml were significantly cytotoxic; the concentration that allowed 10% survival was 120 µg/ml. At 11.6 µg/ml, lindane weakly inhibited metabolic cooperation between 6-TGs and 6-TGr V79 cells. It was concluded from these studies that lindane is not mutagenic in this test system; however, it inhibits metabolic cooperation, mimicking the powerful tumour promotor 12-*O*-tetradecanoylphorbol 13-acetate in this assay system (Tsushimoto et al., 1983).

The morphology of primary monkey kidney cells was examined 24 h after addition of lindane (99.8% in 1% dimethylformamide) to the growth medium, and readings were made daily for three days. Lindane applied at concentrations above 10 mg/litre induced marked cellular damage, and 250 mg/litre had cytotoxic effects (Desi et al., 1977).

7.6.5 Appraisal of mutagenicity and related end-points

The mutagenicity of lindane has been adequately studied. This compound has been extensively investigated for its ability to induce gene mutation in both bacteria and mammalian cells, and for its activity in the assay for sex-linked recessive lethal mutation in *D. melanogaster*. Negative results were obtained consistently. Its ability to induce chromosomal damage and sister chromatid exchange has been investigated in mammalian cells both *in vitro* and *in vivo*, again with negative results. Both assays for DNA damage in bacteria and studies *in vivo* to investigate covalent binding to DNA in the liver of rats and mice following oral administration also gave negative results. The few studies in which positive results were obtained involved invalid study designs or lindane of unknown purity.

Overall, lindane appears not to have mutagenic potential.

Effects on laboratory mammals and in in-vitro test systems

7.7 Carcinogenicity

7.7.1 Mouse

Gamma-HCH was fed to 20 male ICR/JCL mice (five weeks old) at 300 or 600 mg/kg diet for 26 weeks. Increased liver weights were reported in the group receiving the higher dose. Five of 10 mice in this dose group had type 0 or type I liver lesions. Type 0 lesions were characterized as areas of atypical, small liver cells, uniform in size and with a small nucleus, which normally forms round spots and is readily distinguishable from the surrounding liver tissue. Type I lesions were described as 'benign liver tumours' (Goto et al., 1972).

Groups of 20 male mice (eight weeks old) were fed gamma-HCH at 100, 250, or 500 mg/kg diet for 24 weeks. The highest dose level resulted in increased liver weight. No nodular hyperplasia or hepatocellular tumour was observed (Ito et al., 1973b).

Hanada et al. (1973) treated 10–11 dd mice of each sex with lindane at 100, 300, or 600 mg/kg diet for 32 weeks and killed the survivors 5–6 weeks after the end of exposure. Hepatomas were found in 1/3 females and 3/4 males that ingested 600 mg/kg diet and survived for 36–38 weeks; none were found in animals fed 100 or 300 mg/kg diet. At the two higher doses, most animals had atypical proliferations in the liver. Alpha-Fetoprotein could not be identified in the serum of the animals with hepatomas.

In an experiment lasting 110 weeks, 30 male and 30 female CF1 mice were fed lindane (> 99.5%) at 400 mg/kg diet. A group of controls comprising 45 male and 44 female mice were fed a standard diet. Benign and malignant liver tumours were diagnosed in 24% of male controls and 23% of female controls and 93% of treated males and 69% of treated females. Significant mortality (15%) occurred during the early phase of the study in the treated group (Thorpe & Walker, 1973). In a complete reexamination of all slides, the reviewer concluded that lindane had not affected the incidence of hepatocellular carcinomas in animals of either sex but had enhanced the incidence of hepatocellular adenoma (and hyperplastic nodules) in male mice. In this strain, therefore, lindane had a tumorigenic effect only in male mice (Vesselinovitch & Carlborg, 1983).

Herbst et al. (1975) and Weisse & Herbst (1977) studied the carcinogenic potential of lindane at 12.5, 25, or 50 mg/kg diet administered for 80 weeks to 50 male and 50 female Chbi:NMRI mice (a strain with a low (2%) spontaneous rate of hepatomas). The control group consisted of 100 males and 100 females. No evidence of substance-related tumour formation was seen in animals of either sex at any dose level. Electron microscopic examination showed no fine structural hepatocellular alterations.

Groups of 50 B6C3F$_1$ hybrid mice of each sex were fed lindane at 80 or 160 mg/kg diet for 80 weeks and were killed 10–11 weeks after the end of treatment. Hepatocellular carcinomas were found in 5/49 pooled male controls, 2/10 matched male controls, 19/49 males fed 80 mg/kg diet and 9/46 males fed 160 mg/kg diet, and in 2/47 pooled female controls, no matched female controls, 2/47 females fed 80 mg/kg diet, and 3/46 females fed 160 mg/kg diet. Only the incidence of hepatocellular carcinomas in the males at the lower dose was significantly different from that in controls. It was concluded that lindane is not carcinogenic in this test system (US National Cancer Institute, 1977). After a reexamination of the slides, the reviewer was in full agreement with the conclusions of the original authors (Vesselinovitch & Carlborg, 1983).

Wolff & Morrissey (1986) administered diets containing lindane at 160 mg/kg diet for 24 months to three phenotypes of (YS x VY) F$_1$ hybrid mice: obese yellow Avy/a, lean pseudoagouti Avy/a, and lean black a/a. Hepatocellular adenomas were found in 35% of yellow Avy/a mice (9% in controls) and in 12% of pseudoagouti Avy/a mice (5% in controls); no increase in the incidence of liver tumours was seen in the black a/a mice.

7.7.2 Rat

Groups of 10 male and 10 female Wistar rats were fed for life on diets containing 10, 100, or 800 mg/kg diet of powdered lindane or 5, 10, 50, 100, 400, 800, and 1600 mg/kg diet of lindane in corn oil. The life span of the animals was shortened by 20–40% in a dose-dependent manner with administration of 400, 800, and 1600 mg/kg diet, except in those given 800 mg/kg diet of powdered lindane. No increase in tumour incidence was reported in the 200 treated rats (Fitzhugh et al., 1950) (see also section 7.4.1).

In a lifetime study, groups of 10 rats of each sex received lindane at 0, 25, 50, or 100 mg/kg diet. No tumour formation was found (Truhaut, 1954). (Details not given.)

Groups of 18–24 male W rats received lindane (99%) at 500 mg/kg diet for 24 or 48 weeks. High mortality was seen; none of the six or eight surviving animals had developed a liver tumour by 24 or 48 weeks, respectively (Ito et al., 1975).

Groups of 50 Osborne–Mendel rats of each sex were administered lindane for 80 weeks and were then transferred to the control diet for an additional 28–30 weeks; survivors were killed at 108–110 weeks. The males received 320 or 640 mg/kg diet for 38 weeks, lowered thereafter to 160 and 320 mg/kg diet; the females received 320 and 640 mg/kg for two weeks, then 160 and 320 mg/kg for 49 weeks followed by 80 and 160 mg/kg diet for 29 weeks. Matched controls consisted of 10 animals per sex; these were combined for statistical evaluation with 45 untreated male and female rats from other bioassays. No increase in tumour rate was seen in treated groups of either sex (US National Cancer Institute, 1977).

7.7.3 Initiation–promotion

The tumour-initiating activity of gamma-HCH was studied by observing the appearance of phenotypically altered foci in female Wistar rats (Schröter et al., 1987). Groups of 3–8 rats were operated to remove the median and right liver lobes; they were then administered gamma-HCH at 30 mg/kg body weight daily for two weeks, followed by phenobarbital at 50 mg/kg body weight daily for 15 weeks. Liver foci were identified by means of the gamma-glutamyltransferase reaction and morphological alterations. No evidence of initiating activity was found.

Promoting activity was studied by administering *N*-nitrosomorpholine as a single dose of 250 mg/kg body weight by gavage, followed by 4, 15, and 20 weeks' administration of gamma-HCH at 0.1, 0.5, 2.5, 10.0, or 30.0 mg/kg body weight per day. Both the number and the size of altered foci were enhanced by doses of 2–3 mg/kg. The authors concluded that gamma-HCH could be classified as a tumour promotor.

In an experiment using male dd mice (26–30 per group, eight weeks old), administration of Kanechlor-500 at 500 mg/kg diet induced nodular hyperplasia and hepatocellular carcinoma in the livers of mice after 32 weeks' exposure. Administration of lindane (99% pure) at 50, 100, or 250 mg/kg diet with or without the polychlorinated biphenyl at 250 mg/kg diet induced none of those lesions after 24 weeks. Lindane was therefore neither tumorigenic nor a promoter in this experiment (Ito et al., 1973a).

7.7.4 Mode of action

Considerable work has been done using mice generated genetically from (C3H x VY)F_1 or (YS x VY)F_1 mice. The resulting Avy/Avy, Avy/a and A/a crosses contain a genomic locus known as the Agouti locus, which has been linked to tumorigenicity in these mice. Treatment of Agouti mice with lindane at 160 mg/kg diet has been found to saturate the lindane elimination pathways and thereby result in an increased burden of lindane and its metabolites. This excessive build-up could explain the tumorigenicity of lindane, at least when given at 'excessive' levels (Wolff, 1986; Wolff et al., 1986).

The tumour response to lindane has been characterized in (YS x VY) F_1 hybrid mice (Table 12). Lindane increased the incidence of benign tumours only in the Avy/a genotype, while the 'normal' A/a mice had no

Table 12. Carcinogenic responses in normal (A/a), pseudoagouti (A^{vy}/a), and agouti (A^{vy}/a) mice after 24 months of dietary exposure to lindane[a]

Tumour	Phenotype[b]	No lindane		160 ppm lindane		p value
		No.	%	No.	%	
Liver adenoma	B	6/96	6	3/96	3	–
	PS	5/95	5	11/95	12*	0.11
	Y	8/93	9	33/94	35*	8.2E–06
Hepatocellular	B	3/96	3	1/96	1	–
carcinoma	PS	2/95	2	5/95	5	–
	Y	12/93	13**	16/94	27	–
Combined liver	B	9/96	9	4/96	4	–
tumours	PS	7/95	7	16/95	17*	0.036
	Y	20/93	22**	49/94	52*	1.1E–05
Hyperplasia of lung	B	10/96	10	79/96	82*	<E–08
Clara cells	PS	10/95	10	71/94	76*	<E–08
	Y	14/95	15	68/95	72*	<E–08
Lung tumour[c]	B	2/29	2	3/96	3	–
	PS	6/95	6	13/94	14	0.0692
	Y	4/95	4	18/95	19*	0.0012

*Statistically significant dose-related response compared to non-treated comparable controls of the same phenotype
**Liver response increased in obese yellow (A^{vy}/a) mice compared to A/a black controls even in absence of treatment
[a] From Holder & Stöhrer (1989)
[b] B, black normal controls; PS, pseudoagouti; Y, yellow
[c] Not malignant; origin of cells uncertain

tumours. This finding indicates the existence of a genetic predisposing factor, which may be of some importance in evaluating the hazard of exposure to lindane. A phenotypic factor is apparently involved, as mice of the obese yellow phenotype had a greater tumour response in the liver than their isogenic siblings, pseudoagouti mice; such factors may themselves result in more tumours. Tumour incidence was not increased in normal black mice, but the incidence of benign tumours of the liver and lung was increased in Y genotype mice. The time of tumour onset was as early as 18 months in pseudoagouti mice, but the normal black mice had no tumours in the 24-month test period. Avy/a yellow mice thus have a proclivity to form hepatocellular adenomas and lung tumours, which is augmented (and not caused exclusively) by exposure to lindane. The pseudoagouti and normal black mice have a low rate of spontaneous tumours in the liver and lung, but only the pseudoagouti respond to lindane. Thus, some genetically derived mice form benign tumours, but the 'normal' A/a controls do not. Holder & Stöhrer (1989) concluded that these findings are of limited applicability to the situation in humans.

Tumour promotion was tested as a mechanism of action for lindane in both the yellow and pseudoagouti variants of the (C3H x VY) F_1 mouse using phenobarbital, which increases the incidence of benign tumours in the liver of yellow mice, as the tumour-promoter. Lindane at a dose of 160 mg/kg diet may exceed saturation of the metabolic mechanisms, especially in yellow mice with the (C3H x VY)F_1 genotype. Chadwick et al. (1987) found reduced elimination of lindane in yellow and pseudoagouti mice and explained their findings as follows: The yellow mouse carrying the Avy locus has a propensity for tumorigenicity, which is enhanced by the yellow obese phenotype. The lungs and livers of these animals therefore are very likely to contain cells that are already transformed, whereas in normal black mice there may be none or very few transformed cells. Hence, Avy strain mice would be expected to respond to a tumour promoter, whereas black mice would not; this was also the pattern of tumour response observed. The authors concluded that, because phenobarbital promotes tumours in this strain of mice, lindane is also a promoter.

Oesch et al. (1982) studied the specific activities in CF_1 and $B6C3F_1$ mice and Osborne–Mendel rats of some of the enzymes thought to be involved in lindane metabolism. Lindane was administered at 51–360 mg/kg diet for three days or three months. No clear change was seen in animals treated for three days, but changes in enzyme activity were noted after three months' treatment. In the CF_1 strain (sensitive to liver tumour induction), a large increase in liver weight was observed; this was not the case in $B6C3F_1$ mice. In the Osborne–Mendel rats, a smaller increase was found.

Glutathione-S-transferase activity was increased in CF_1 mice and to a lesser extent in $B6C3F_1$ mice and the rats. Increased glutathione-S-transferase activity may lead to rapid conjugation of glutathione with reactive metabolites, as, for example, epoxides derived from lindane. Rat liver microsomes had more UDP-glucuronosyltransferase activity than those from mouse liver. This increased activity in rats could also lead to rapid conjugation of phenols derived from lindane. The most striking difference, however, was that CF_1 mice had more monooxygenase activity and less epoxide hydroxylase activity than rats; whether either of these changes would result in an accumulation of reactive epoxides from lindane remains to be elucidated.

Iverson et al. (1984) studied the ability of ^{14}C-gamma-HCH to bind to liver macromolecules of untreated and phenobarbital-pretreated male HPB black mice *in vivo* and *in vitro*. There was preferential binding of gamma-HCH to protein but not to DNA.

These studies in mice indicate that lindane does not behave as an initiator, in that it does not induce the preneoplastic foci seen with known carcinogens, such as *N*-nitrosomorpholine and *N*-methyl-*N*-nitrosourea (Holder & Stöhrer, 1989). Lindane can, however, act as a tumour promoter, in that it caused outgrowth of foci and increased the areas of the foci, indicative of preneoplastic conditions. Whether these foci actually go on to form tumours was not determined.

The notion that lindane has some characteristics in common with tumour promoters is corroborated by the finding that it inhibits cell-to-cell communication of the low-molecular-weight compound, tritiated uridine. Trosko (1982) found that such inhibition occurred when cells were pretreated with a variety of free-radical scavengers and suggested that the inhibition might involve a free-radical generating process. Some tumour promoters have been suggested to act by a mechanism involving free radicals (Kensler & Trush, 1984; Rao & Reddy, 1987).

If lindane acts by the tumour promotion mechanism suggested by formation of gamma-glutamyltransferase-positive foci in the liver and inhibition of cell-to-cell communication, it is likely to be a dose-rate-limited process because of its known reversibility. That is, the compound must be administered at above a certain amount and rate or its carcinogenic effects are reversible and cease to be manifested. Such a mechanism would therefore probably result in a sigmoid response in models.

Zeilmaker & Yamasaki (1986) studied the effect of lindane on gap-junctional intercellular communication in cultured Chinese hamster V79

Effects on laboratory mammals and in in-vitro *test systems*

cells, grown as a monolayer using a microinjection/dye transfer technique. Intercellular communication via gap junctions is thought to play a crucial role in cell proliferation and differentiation and in tissue homeostasis, and consequently in carcinogenesis. Lindane inhibited junctional communication in a dose-response relationship (0–20 µg/ml) after a 60-h exposure, but inhibition was seen after 24 h incubation only with the highest dose level. In an earlier study, lindane strongly inhibited metabolic cooperation between V79 cells at a non-toxic dose of 10 µg/ml (Tsushimoto et al., 1983).

Another explanation for the finding that Avy/a yellow mice have a predisposition for tumorigenicity but not necessarily for carcinogenicity is their reduced immunocompetence, as evidenced by decreased antibody response to T-cell-dependent immunogen tetanus toxoid, enhanced antibody response to T-cell-independent immunogen type III pneumococcal polysaccharide, decreased rates of carbon clearance and increased levels of immunoglobulin A. The pseudoagouti mice did not have reduced immunocompetence and had reactions similar to those of normal black A/a mice in these immunological tests (Holder & Stöhrer, 1989).

The lindane metabolite, 2,4,6-TCP, constitutes a significant proportion of the urinary metabolites of lindane and is considered to be a carcinogen. However, direct measurements of comparative potency indicate that TCP contributes only a small fraction of the 'lindane cancer potency' and therefore may not add significantly to the quantitative impact of lindane. The notion that TCP adds quantitatively to the carcinogenicity of lindane *per se* remains a major element in the evaluation of the carcinogenic hazard of lindane to humans (Holder & Stöhrer, 1989).

7.7.5 Appraisal of carcinogenicity

Studies to define the carcinogenic potential of lindane have been conducted with mice and rats, at doses of up to 600 mg/kg diet in mice and up to 1600 mg/kg diet in rats. In some studies, the dose levels exceeded the maximum tolerated dose. Hyperplastic nodules and/or hepatocellular adenomas were found in studies with mice at doses from 160 mg/kg diet. Two studies using mice and one study using rats, with dose levels of up to 160 mg/kg diet in mice and 640 mg/kg diet in rats, showed no increase in the incidence of tumours. The results of studies on initiation–promotion, on mode of action, and on mutagenicity indicate that the tumorigenic effect of gamma-HCH in mice results from non-genetic mechanisms.

7.8 Special studies

7.8.1 Immunosuppression

Desi (1976) and Desi et al. (1978) reported the results of a subacute study in which groups of 30–36 male rabbits were treated orally, five times per week for 5–6 weeks, with doses of lindane representing 0, 1/5, 1/10, 1/20, and 1/40 of the oral LD_{50}, which was 60 mg/kg body weight. Once a week, different doses of *S. typhimurium* 'Ty 2' vaccine were injected intravenously. The humoral immune response was determined by the tube agglutination test. A linear regression was found between the dose of lindane and reduction in antibody titres, in a time-dependent manner. The lowest dose, 1.5 mg/kg body weight (1/40 of the oral LD_{50}) caused no immunosuppression.

7.8.2 Behavioural studies

The learning rate in a maze and responses to conditioning in a Skinner box were studied after feeding lindane at daily doses of 2.5, 5, 10, or 50 mg/kg body weight to Wistar rats for 40 days. In the maze, no effect was seen with 2.5 mg/kg, but at 5 mg/kg there was stimulation accompanied by an increased error rate in maze running activity; at 10 and 50 mg/kg, the animals became sedated and committed more errors than the controls. In the Skinner box, stimulation was seen with 2.5 and 5 mg/kg. At 10 mg/kg, no difference was seen from the controls, whereas animals treated at 50 mg/kg were less active than the controls (Desi, 1974).

7.8.3 Neurotoxicity

Lindane produces a variety of neurological effects, both central and peripheral, in mammals. The induced increase in neuronal excitability and the underlying mechanisms of action have been investigated both *in vivo* and *in vitro*.

7.8.3.1 Dose-response studies using intact animals

The effects of lindane on body temperature, food intake, and body weight were studied in Wistar rats given single or repeated non-convulsant oral doses. Groups of eight male and eight female rats were given lindane as a single oral dose of 30 mg/kg in olive oil. Controls received olive oil

Effects on laboratory mammals and in in-vitro test systems

alone. Further groups of eight males and eight females received 10 mg/kg and two groups of male rats received 30 mg/kg once daily for seven days at either thermoneutral ambient temperature or cold ambient temperature (4 °C). The single dose of 30 mg/kg significantly decreased core temperature 5 h later; this lindane-induced hypothermia was strongly potentiated by cold stress in rats kept at 4 °C. A decrease in body weight gain was also observed. No hypothermic effect was seen with 10 mg/kg (Camon et al., 1988a).

The relationship between the brain concentration of lindane and its convulsant effect was studied in male Wistar rats administered lindane (99.5%) dissolved in olive oil daily by gavage at doses of 5, 12, or 20 mg/kg body weight for 12 days. The mean plateau concentration in brain was achieved within 5–8 days. There was a strong correlation between the doses administered and the concentration in brain at the plateau. A convulsant response was not seen with 5 mg/kg, but tonic convulsions occurred at the two higher doses. The rate of response (percentage of rats with convulsions) was also correlated with the log of the concentration of lindane in brain. The concentration in brain decreased after 12 days of daily administration of doses of 5 and 12 mg/kg, but not with 20 mg/kg (Tusell et al., 1988).

Camon et al. (1988b) investigated the effect of convulsant and non-convulsant doses of lindane on regional glucose uptake in the brain. Male Wistar rats received intraperitoneal injections of ^{3}H-2-deoxyglucose, and the amount of label in different brain structures was assayed by liquid-scintillation counting in 18 dissected brain regions. Lindane at a single convulsant dose (150 mg/kg orally) increased 2-deoxyglucose uptake in olfactory tubercules, hypothalamus, hippocampus, paraflocculi, and the post-medulla. With a single, non-convulsant dose of 30 mg/kg, the uptake of 2-deoxyglucose was less affected; after treatment with 10 mg/kg per day for one week, 2-deoxyglucose uptake was observed in superior colliculi but was decreased in the parietal cortex. The increased uptake in limbic regions seen at the convulsive dose correlates with the experimentally observed association between signs of poisoning induced by lindane and damage to the limbic system.

Intraperitoneal injections of gamma-HCH (99.0%) in corn oil at 80–480 mg/kg body weight increased the accumulation of cerebellar cyclic GMP in male CD-1 mice. Furthermore, it inhibited the binding of ^{3}H-*tert*-butylbicyclo-*ortho*-benzoate (a ligand for the GABA-A receptor-linked chloride channel) in mouse cerebellum (Fishman & Gianutsos, 1987).

Fishman & Gianutsos (1988) gave male CD-1 mice gamma-HCH at single intraperitoneal doses of 80–400 mg/kg body weight in corn oil. At the lowest dose, gamma-HCH increased the lethality and the frequency of tonic/clonic seizures induced by intraperitoneal injection of 50 mg/kg pentylenetetrazole or 20 mg/kg picrotoxin but had no effect on locomotor activity.

Sunol et al. (1988) studied the effect of administering lindane by gavage at 150 mg/kg body weight in olive oil on the GABAergic and dopaminergic systems, by measuring the concentrations of GABA, dopamine and its metabolites in seven brain areas at the onset of seizures. All animals suffered tonic convulsions 18.3 ± 1.4 min after lindane administration. The concentration of GABA was decreased only in the colloculi and not in the other areas. Dopamine concentrations were increased in the mesencephalon, and those of its metabolite, dihydroxy-phenylacetic acid, were also increased in the mesencephalon and the striatum (abstract only).

In studies by Desi (1983), adult female CFY rats were given a daily dose of lindane (99.5%) at 2.5 mg/kg body weight. This dose level had no functional, neurological, electroencephalographic, or psychophysiological effect, used as early signs of disturbances of the nervous system. A dose of 5.0 mg/kg body weight altered the electrical activity of the brain, as indicated by changes in the complex electroencephalograph, the number of changing electroencepahlographic bands and the index number. In behavioural experiments, running speed and number of errors indicated an inhibitory effect of lindane at 5.0 mg/kg body weight on learning capacity; this result was not seen with 2.5 mg/kg.

Müller et al. (1981) studied the electroneurophysiological effects of various HCH isomers on groups of 15 male Wistar rats by feeding them diets containing each isomer for 30 days. Conduction velocity delay was observed in the animals fed the gamma isomer at a daily dose of 25.4 mg/kg, but not at 12.3 or 1.3 mg/kg. The greatest delay was induced by the lindane metabolite gamma-PCCH (38–783 mg/kg body weight).

Lindane was reported to lower the threshold for kindled seizures (resulting from repetitive stimulation of the limbic system within the brain) in rats (Joy et al., 1982, 1983). In these experiments, stimulating electrodes were implanted in the amygdala and other part of the limbic system, and the animals were stimulated with a 1-s train of pulses at 60 Hz on each day of the study. In this procedure, no response is elicited initially; however, with repeated stimulation, the procedure induces increasing levels of electrical seizure, with clonic seizures resulting after many trials. A developing after-

Effects on laboratory mammals and in in-vitro test systems

discharge becomes progressively longer, and the severity of the accompanying motor signs becomes more pronounced, until kindling is completed. The subject then exhibits stable convulsive responses for weeks or months afterwards. The duration of the electrical seizure and the severity of the behavioural response were found to increase much more rapidly when lindane was administered at daily oral doses of 1, 3, or 10 mg/kg body weight 3 h before each kindling trial. These effects were found to be dose-dependent, and a threshold exposure of 0.5 mg/kg per day was calculated. Rats administered lindane at this concentration were found to develop brain levels of lindane which fluctuated between 0.2 and 0.4 µg/g.

Joy & Albertson (1987a,b) demonstrated that lindane alters dentate gyrus granule response to perforant path input in the intact rat in a manner indistinguishable from picrotoxin or pentylenetetrazol, which are known GABA-mediated chloride channel antagonists. In this study, 19 male Sprague–Dawley rats were anaesthetized with urethane, electrodes for stimulating and recording responses from the dentate gyrus of the hippocampus were implanted, and the animals were placed in a stereotaxic device. Lindane was then administered intraperitoneally in dimethyl-sulfoxide to each animal at sequential doses of 5, 10, 20, and 40 mg/kg body weight. Single or paired electrical stimuli were presented at different intensities and at different intervals to evaluate the effects of lindane on inhibition and facilitation. These studies demonstrated a dose-dependent change in perforant path granule cell function, manifested as an increase in the excitability of the granule cell to other stimuli. Lindane was also found to induce a small but statistically significant, dose-dependent increase in presynaptic inhibition, as well as a significant increase in postsynaptic inhibition. A dose-dependent effect on GABA-mediated inhibition was measurable at exposures that were not convulsant in unanaesthetized animals. The results of this *in-vivo* study indicate that inhibition of GABA-mediated chloride channels in the brain is probably an important mechanism by which lindane produces neuronal hyperexcitability and convulsions.

7.8.3.2 Studies on mechanism

Although the precise mechanism by which lindane exerts its neurotoxic action is not fully resolved, studies using preparations of synaptosomes (pinched-off nerve endings) and of cholinergic neuromuscular junctions, as well as studies using intact animals, have provided insight into this issue. The results of representative studies with each type of preparation are summarized in Table 13.

Table 13. Known effects of lindane on central and peripheral nerves or muscle tissue

Effect	Species	Type of preparation	Reference
Inhibition of GABA-mediated chloride ion uptake in CNS	Rat	in vivo	Joy & Albertson (1987a,b)
	Rat	Brain synaptosomes/ neuronal membrane	Eldefrawi et al. (1985) Matsumura &Tanaka (1984) Abalis et al. (1985, 1986)
	Mouse	Brain synaptosomes	Fishman & Gianutsos (1988)
Increased availability of intracellular Ca^{++}	Rat	Brain synaptosomes	Hawkinson et al. (1989)
	Rat	Neuroblastoma cells	Joy et al. (1987)
	Rat	Neurohybridoma cells	Joy et al. (1988)
	Frog	Neuromuscular junction	Joy et al. (1987) Vogel et al. (1985) Publicover & Duncan (1979)
Decreased conduction velocity	Rat	Tail nerve	Müller et al. (1981)
Mitochondrial damage	Frog	Skeletal muscle	Publicover et al. (1979)
Na^+-K^+-ATPase inhibition	Frog	Skeletal muscle	Pandy et al. (1985)
	Rat	Liver mitochondria	Srinivasan & Radhakrishnamurty (1988)
Ca^{++}-Mg^{++}-ATPase inhibition	Beef	Brain	Uchida et al. (1974)
	Rat	Liver mitochondria	Srinivasan & Radhakrishnamurty (1988)

Lindane was shown to inhibit the uptake of chloride ions at inhibitory synapses in the brain (Matsumura & Tanaka, 1984; Abalis et al., 1985, 1986; Fishman & Gianutsos, 1988), and it is this mode of action that is now widely considered to account primarily for the convulsant activity of this insecticide. Because of its structural similarity to picrotoxinin, lindane has a good geometric fit to the picrotoxinin-binding site at the outer end of the chloride channel. Once bound, lindane is believed to block the action of the neurotransmitter GABA, which mediates the entry of the Cl⁻ necessary for inhibitory neuronal function (Matsumura, 1985). More recent studies by Joy & Albertson (1987a,b) provide evidence of such inhibition of GABA in rats *in vivo*, demonstrating that: (a) this mechanism operates at clinically relevant exposure levels (5–40 mg/kg); (b) the magnitude of this effect is dose-dependent; and (c) this effect can be clearly measured at subconvulsant exposure levels in unanaesthetized subjects.

Effects on laboratory mammals and in in-vitro test systems

Lindane has also been demonstrated to increase the excitability of presynaptic cholinergic neurons at central (Hawkinson et al., 1989) and peripheral (Vogel et al., 1985; Publicover & Duncan, 1979) synapses. Using a nerve–muscle preparation, a universally accepted model for a cholinergic synapse, Publicover & Duncan demonstrated that concentrations of lindane as low as 5×10^{-5} mol/litre increase the spontaneous release of acetylcholine. Similar results were reported by Vogel et al. (1985), who observed increases up to 100 fold in spontaneous release of neurotransmitter in the presence of lindane at 10^{-4} mol/litre. This increase was found in both studies to be somewhat dependent upon the concentration of extracellular calcium ions. In the presence of normal (~1–2 mmol/litre) Ca^{++}, the increase in spontaneous transmitter release was 2.5–20 times greater than that measured when the extracellular Ca^{++} level was buffered to 10^{-7} mol/litre, the homeostatically maintained intracellular Ca^{++} concentration. Hawkinson et al. (1989) verified the involvement of Ca^{++} in the enhanced release of acetylcholine using a rat brain synaptosome preparation, and he postulated that the increased excitability of cholinergic synapses (widely dispersed throughout the brain) may contribute to the convulsant effect of lindane.

Pandy et al. (1985) reported that lindane weakly inhibits Na^+-K^+-ATPase activity in skeletal muscle, and they suggested that lindane can inhibit Ca^{++}-Mg^{++}, Ca^{++}-, and Mg^{++}-ATPase. Uchida et al. (1974) reported decreased Ca^{++}-Mg^{++}-ATPase activity in beef brain. Since all of these ATPases are involved in the maintenance of intracellular Ca^{++} concentration and have been identified in a neuronal membrane preparation (Yamaguchi et al., 1979), it is possible that the disruption of homeostatic calcium regulation by lindane contributes to its excitatory action on the central nervous system.

7.8.3.3 Summary

Chronic exposure to low levels of lindane can result in proconvulsant activity, as demonstrated experimentally using the kindling model of experimental epilepsy. Lindane has also been shown to cause convulsions in rats administered oral doses of 12 mg/kg daily for 12 days; a dose of 5 mg/kg body weight did not induce convulsions within that period. The convulsive effect has been suggested to be associated with inhibition of GABA-mediated chloride channels in the brain, as demonstrated experimentally in mammals both *in vivo* and *in vitro*. Changes in the electroencephalogram and decrements in a variety of behavioural parameters have been observed with lindane at a dose of 5 mg/kg body weight per day for 40 days, but not at 2.5 mg/kg for 22 days. A delay in peripheral nerve conduction velocity was observed in rats administered a dietary concentrations of 25 mg/kg body weight, but not at 12 mg/kg.

'.9 Factors that modify toxicity; toxicity of metabolites

Pretreatment of rats with lindane minimized or inhibited the convulsive effects of pentazole, picrotoxin, loramine, strychnine, mintracol, cyclohexane sulfonamide and electro-shock (Herken, 1950a,b,c, 1951; Coper et al., 1951; Kewitz et al., 1952; Lange, 1965). It was demonstrated that premedication for two weeks with daily doses of 2 mg lindane not only accelerated the metabolism of other chemicals, but also caused an acceleration of its own breakdown: in Fischer rats given two oral administrations of lindane in oil at 2 mg/kg, there was increased excretion of glucuronic acid conjugates (Chadwick et al., 1971).

The smallest single oral dose of lindane that reduced pentobarbital sleeping time in FW-49 rats was 5 mg/kg body weight (Schwabe & Wendling, 1967). The smallest single intraperitoneal dose of lindane that shortened hexabarbital anaesthesia was 15 mg/kg body weight. A similar result was obtained in Sprague–Dawley rats fed a diet containing lindane at 0.5 mg/kg; the effect was more distinct with a dose of 4 mg/kg diet (Kolmodin-Hedman et al., 1971).

8. EFFECTS ON HUMANS

8.1 Exposure of the general population

8.1.1 *Acute toxicity, poisoning incidents*

Several cases of fatal poisoning and numerous cases of non-fatal illness caused by or ascribed to lindane have been reported. These incidents were either accidental, intentional (suicide) or due to gross neglect of safety precautions. In many of these cases, the effects ascribed to lindane were more likely to have been due, in total or in part, to other substances. A critical review of these cases is provided by Hayes (1982).

The toxic or lethal dose appears to vary considerably with the carrier and/or the degree of homogenization of the product. Under certain conditions, 10–20 mg/kg body weight can present a lethal hazard to humans, but higher concentrations can be tolerated when followed by timely and appropriate medication. Starr & Clifford (1972) described a case of acute lindane intoxication in a 2.5-year-old child who had severe epileptiform convulsions after ingesting presumably two 0.78-g pellets containing 95% lindane. The child recovered under medical care.

A suicide attempt was reported by Ohly (1973). The person ingested about 100 ml of a 25.5% lindane emulsion concentrate, which corresponds to a dose of about 309 mg/kg body weight. No vomiting or diarrhoea was seen, but severe convulsions occurred within the first 24 h after lindane ingestion. Elevated serum levels of glutamic-oxaloacetic transaminase, lactic dehydrogenase, glutamic-pyruvate transaminase, and creatine phosphokinase activity, in conjunction with the results of a liver biopsy, suggested fatty degeneration and severe toxic damage to the liver. After five weeks of clinical treatment, the patient recovered completely.

A number of cases of poisoning cases have been described after medical treatment or abuse (Davies et al., 1983; Kurt et al., 1986; Petring et al., 1986; Berry et al., 1987).

Clinical signs of intoxication can appear from a few minutes to some hours after intake of lindane, depending on the route of administration, the formulation, the concentration of lindane, and the quantity involved. In mild cases, indisposition, nausea, dizziness, restlessness, frontal headaches, and sometimes vomiting occur. Muscular fasciculation, disturbances of equilibrium, ataxia, and tremor may appear. Pains in the upper abdomen are frequently coupled with diarrhoea and uncontrolled micturition. Clonic-tonic convulsions of some minutes' duration may occur, and these may recur after several hours or even days in response to optical, tactile, and acoustic stimuli. In fatal cases, death follows several hours to several days after intake. The cause of death is usually central respiratory failure or acute circulatory collapse, often after convulsions (Hayes, 1982; Jaeger et al., 1984).

1.2 Effects of short- and long-term exposures—controlled human studies

1.2.1 Oral administration

Lindane given orally as a vermicide at a dose of 45 mg to an adult patient (26 years old) in poor condition induced convulsions, nausea, and vomiting. Recovery took place within 3–4 h. Of 15 patients in the same trial, who repeatedly received up to 30 mg/person for up to three days, six complained of nausea; no other symptom was reported (Graeve & Herrnring, 1949).

Severe toxic symptoms were described in healthy volunteers after oral intake of 15–17 mg/kg body weight of lindane in a liquid carrier (Hofer, 1953; Schmiedeberg & Wasserburger, 1953). Reports of toxic effects after administration of lindane against scabies indicate that children are more sensitive to lindane than adults, and rare cases of aplastic anaemia have been reported (G. Volans, National Poisons Unit, London, letter to IPCS, 1989).

1.2.2 Dermal application

Clinical reports after pharmaceutical use of lindane cutaneously suggest that exposures somewhat higher than 5 mg/kg body weight per day do not usually result in acute neurotoxic symptoms. No cause–effect relationship was found between lindane and blood dyscrasias (Ginsburg et al., 1977; Kramer et al., 1980; Morgan et al., 1980).

Three groups, each consisting of one male and one female volunteer, received an application of 30 g of a 0.3% commercial lindane emulsion over the entire body (except the head and the angle of the elbow) on three consecutive days. The first group removed the emulsion by washing with soap and water 3 h after application; the second removed it 10 h after

Effects on humans

application by washing with water only at body temperature, and the third group by washing with soap and water 10 h after application. A group of scabies patients were given an application of 50 g of the emulsion over the entire body, except the angles of the elbows. Blood samples were taken regularly from all participants in the trial over a period of 1–144 h. The highest average serum concentration found approximately 5.5 h after application in the healthy volunteers was slightly less than 5 µg/litre, while in scabies patients an average of about 200 µg/litre was found after 4 h. It was concluded that scabies patients deposit greater concentrations of lindane in their bodies than do healthy persons; the levels were higher in women than in men (Lange et al., 1981; Zesch et al., 1982).

Feldmann & Maibach (1974) studied the absorption and excretion of ^{14}C-labelled lindane dissolved in acetone, applied at 4 µg/cm^2 to the forearms of six human subjects. The usual dose of radioactivity was 1, 2, or 5 µCi. Data obtained after intravenous dosing was used to correct the results obtained on skin penetration for incomplete urinary recovery. Total excretion of ^{14}C after topical application was 9.3% ± 3.7 of the dose in five days. Skin absorption is incomplete because the chemical is lost from the skin surface by washing, evaporation or gradual exfoliation of outer layers of the stratum corneum. The amount absorbed into the body depends on the relationship between the speed with which it penetrates the skin and the speed with which it is lost from the skin surface.

Serum concentrations of gamma-HCH were determined in nine children (3.5–18 years old) following application of a 1% gamma-HCH shampoo to treat pediculosis capitis (head lice). The shampoo was applied vigorously to dry hair in a sufficient amount to saturate the hair and scalp thoroughly. After 10 min, small quantities of water were added until a lather formed, and shampooing was continued for 4 min; thereafter, the hair was rinsed and blown dry. Gamma-HCH was present in the serum of all children 2–24 h following the application. A maximal concentration of 1.4 µg/litre was found after 2–4 h; this level decreased within 24 h to 0.41 µg/litre. Re-treatment increased the maximal level to 3.6 µg/litre (Ginsburg & Lowry, 1983).

Nitsche et al. (1985) applied emulsions of 0.3 and 1.0% ^{14}C-lindane to defined areas of 2 cm^2 intact or stripped skin. They found that the flux of lindane in the skin was time-dependent: Generally, the concentration increased with the depth of the layer. Increased availability of lindane, induced by absence of the stratum corneum and a long application period, resulted in preferential accumulation in the epidermis, with none in the subcutaneous fat. When intact skin was washed with soap and water 3 h after the application, the concentration of lindane in the layer below the stratum corneum 7 h later was higher than that in skin that had not been washed. Washing of intact skin after a short penetration period (3 h) resulted in introduction of lindane; this phenomenon was not seen with stripped skin. Lindane could be more effectively removed from the stratum corneum with soap and water than with water alone.

1.1.3 Epidemiological studies (general population)

The rate of mortality from liver cancer in the USA was related to the 'domestic disappearance' of organochlorine pesticides. In 1962, 18 and 15 years after the introduction of DDT and technical HCH, respectively, by which time any increase in the mortality rate from primary liver cancer would be manifest, the number of cases of primary liver cancer as a percentage of the total number of deaths from liver cancer began a gradual, steady decline, from 61.3% in 1962 to 56.9% in 1972. The death rate (per 100 000 per year) from primary liver cancer during this period declined from 3.46 to 3.18 (Deichmann & MacDonald, 1977).

A considerable number of case reports have been published in which different blood dyscrasias were described in people who had been exposed to lindane or to lindane and other chemicals (Mendeloff & Smith, 1955; Albahary et al., 1957; Jedlicka et al., 1958; Stieglitz et al., 1967; West, 1967; Hoshizaki et al., 1969; Vodopick, 1975). The issuance by the US Environmental Protection Agency (1977) of a 'rebuttable presumption against registration and continued reregistration' of lindane in 1977 was triggered in part by the problem of blood dyscrasias. Studies conducted over periods of several weeks to several years, however, have given no indication that there might be a cause–effect relationship between exposure to lindane and blood dyscrasias (Milby & Samuels, 1971; Samuels & Milby, 1971; Morgan et al., 1980; Wang & Grufferman, 1981), and this was the final conclusion of the US Environmental Protection Agency (1984a).

1.2 Occupational exposure

1.2.1 Toxic effects

Evaluation of the effects of gamma-HCH in occupationally exposed workers is seriously hampered by the fact that most of the studies are of workers who were exposed during manufacturing and handling of lindane, or in the handling or spraying of technical-grade HCH among other

Effects on humans

pesticides. All of these groups are also potentially exposed to other HCH isomers, to impurities and to other (process) chemicals. It is, therefore, difficult to relate the effects found in these studies to any individual substance. The only such studies mentioned here are those that were considered to be useful for the evaluation.

Kolmodin-Hedman (1974) investigated blood levels of gamma-HCH in 54 spraymen exposed to 4% lindane and other insecticides in the form of aerosols and mists and who also occasionally diluted stock solutions. Their exposure varied from once daily to once weekly, and the length of exposure was 1–20 years; they did not always wear protective gloves and respirators. Spraymen exposed to lindane had mean plasma levels of gamma-HCH of 6.4, 7.5, and 9.9 µg/litre; a maximal concentration of 87.0 µg/litre was found. Antipyrine half-lives in exposed subjects and in controls were compared to investigate whether lindane induces drug metabolism in humans. In 26 workers exposed mainly to lindane, the mean antipyrine plasma half-life was significantly shorter than that in 33 controls: 7.7 $\pm$ 2.6 h compared with 13.1 $\pm$ 7.5 h, respectively. Induction occurred with plasma levels above 10 g/litre; workers who were exposed to lindane had shorter antipyrine and phenylbutazone half-lives when their serum levels were above this value. Hyperlipoproteinaemia (defined as a serum cholesterol level above 800 mg/litre and a phospholipid level above 1400 mg/litre in the HDL fraction of the serum lipoproteins) was found in 40% of the spraymen (Kolmodin-Hedman, 1974, 1984).

Three workers mixed rapeseed manually with 75% lindane powder, which also contained 10% thiram, and usually closed the sacks by hand. This mixing procedure was repeated up to 80 times during a workshift. The working period comprised the spring months of each year, and the total exposure period varied between one and five years. Gloves and masks were not always used, so dermal and respiratory exposures were intensive. The plasma levels during exposure in the three people who prepared rapeseed were 102, 100, and 4.2 µg/litre; the last person had less frequent exposure than the other two. The plasma level in the person who had 100 µg/litre during exposure had decreased to the level before exposure within five months (Kolmodin-Hedman, 1974, 1984).

Workers engaged in the production of lindane and exposed for at least six months (8 h/day; wearing face masks in an air-ventilated location) were tested for the presence of chromatid-type and labile chromosome-type aberrations in their lymphocytes. The frequency of stable chromosomal aberrations did not differ significantly from that in normal controls (Kiraly et al., 1979).

Herbst (1976) examined workers engaged in the production of lindane in three factories. The people were exposed not only to the gamma isomer but also to the alpha, beta, delta, and epsilon isomers. The average length of service was slightly more than 10 years. Of the 118 persons examined, 115 were men and three were women, and the average age was 39 years. No abnormality was detected in the haematopoietic system, the liver, the kidneys, or the nervous system.

A series of reports were made on groups of 54–60 male workers (24–62 years of age) in a lindane-producing factory, with a geometric mean duration of exposure of 7.2 years (range, 1–30 years) (Baumann et al., 1981; Brassow et al., 1981; Tomczak et al., 1981). The lindane concentrations in serum were in the range 5–188 µg/litre; that of alpha-HCH was 10–273 µg/litre, and that of beta-HCH, 17–760 µg/litre. None of the controls had a HCH concentration in serum above the limit of detection (0.7 µg/litre). The time-weighted average threshold limit value of 0.5 mg/m^3 was not exceeded at any of the workplaces (range, 0.004–0.15 mg/m^3); the level of alpha-HCH was 0.002–1.99 mg/m^3 and that of beta-HCH, 0.001–0.38 mg/m^3. Only small deviations were found in some laboratory tests: higher polymorphonuclear leukocyte count, lower lymphocyte count, higher reticulocyte count, lower prothrombin level, lower blood concentrations of creatinine and uric acid; these findings were regarded as of no pathological significance. No significant difference was observed in total red cell, white cell, and platelet counts, in haemoglobin content, or in levels of gamma-glutamine transferase, glutamic-oxaloacetic transaminase, glutamic-pyruvic transaminase, lactic dehydrogenase, cholinesterase, triglyceride, cholesterol or urea. No sign of health impairment was observed. Examination of reflexes, sensitivity, amplitude and frequency of fore-finger tremor and manual skills showed no significant difference between the HCH-exposed and the control groups. In addition, no pathological result was obtained in tests by electromyography, for maximal motor nerve conduction velocity in ulnar nerves, or for neuromuscular conduction. Furthermore, electroencephalographic recordings showed no specific pathological sign. The authors concluded that, even after decades of occupational exposure to HCH, no sign of neurological impairment or perturbation of neuromuscular function had occurred. Serum levels of luteinizing hormone were higher in HCH-exposed workers than in controls (8.8 vs 5.7 mIU/ml). Levels of follicle stimulating hormone were insignificantly higher and testosterone levels were insignificantly lower in exposed men than in controls.

In malaria-control workers, who sprayed technical-grade HCH for 16 weeks, the serum level of gamma-HCH increased from a mean of 0.009 to 0.037 mg/litre in previously unexposed workers, and from 0.009 to

Effects on humans

0.034 mg/litre in workers who had been exposed during three previous spraying seasons (Gupta et al., 1982). In comparison with the alpha and beta isomers, the gamma isomer was the least cumulative and the least persistent in serum. These findings, as well as those of Milby et al. (1968), suggest that gamma-HCH levels in blood are mainly a reflection of recent exposure to lindane.

Nigam et al. (1986) studied 64 employees at a manufacturing plant who were directly or indirectly associated with the production of HCH. The exposed group comprised 19 workers who handled and packaged the insecticide, 26 plant operators and supervisors who were exposed indirectly to HCH, and 19 members of the maintenance staff who visited the plant frequently. The control group consisted of 14 workers who had no occupational contact with HCH. The length of exposure varied from 0 to 30 years. The mean serum concentrations of lindane in the four groups were: control, 0.0007, maintenance staff, 0.0227, indirect exposure, 0.016, and handlers, 0.0571 mg/litre. Alpha-, beta-, and delta-HCH were also present; the total HCH concentrations in serum were 0.0514 mg/litre in the controls, 0.1436 mg/litre in the maintenance staff, 0.2656 mg/litre in the indirectly exposed workers, and 0.604 mg/litre in the handlers. Most of the directly and indirectly exposed workers had paraesthesia of the face and extremities, headache, and giddiness, and some had symptoms of malaise, vomiting, tremors, apprehension, confusion, loss of sleep, impaired memory and loss of libido. The same symptoms were found in the group of maintenance workers but were less severe and occurred in fewer cases.

Chattopadhyay et al. (1988) studied 45 male workers exposed to HCH during its manufacture and compared them with 22 matched controls. Paraesthesia of the face and extremities, headache, giddiness, vomiting, apprehension, and loss of sleep, as well as some changes in liver function were reported. These changes were found to be more closely related to intensity of exposure (as measured by HCH levels in blood serum) than to duration of exposure. The measured exposures to total HCH were 13 to 20 times higher than those in the control groups (details not given). Of the total HCH, 60–80% was gamma-HCH.

Plasma levels of gamma-HCH and urinary levels of three TCPs were measured in 45 forestry workers who were engaged in dipping conifer seedlings in a gamma-HCH solution, transporting the dipped seedlings to planting sites or planting the seedlings. Protective clothing was supplied. The work started in April, and until June the workers had plasma concentrations below the detection limit (5 nmol/litre). From June onwards, there was an upward trend in the number of workers who had gamma-HCH levels in plasma of up to 40 nmol/litre. In July, levels of 59, 75, and 123 nmol/litre were found in three workers, and the latter two had symptoms of poisoning that included feeling unwell with a flu-like illness, fatigue, sore throat, and nausea. No sign of hepatotoxicity was observed. People with levels greater than 70 nmol/litre were removed from further exposure. When exposure ceased at the end of July and the people who had had elevated levels were monitored in August, 80% of the workers had no detectable gamma-HCH in their plasma; the group mean concentration was 16 nmol/litre. By September, all plasma levels had returned to the pre-exposure level, below the detection limit. The mean half-life of gamma-HCH in plasma was calculated to be about eight days. 2,4,6-, 2,3,5-, and 2,4,5-TCP were the major metabolites of gamma-HCH (Drummond et al., 1988).

Neurological studies on 37 workers exposed to lindane over a period of two years revealed three with serious electroencephalographic disturbances; minor symptoms and signs were seen in 14 of the workers. No change was observed in the electroencephalographic patterns of 21 of the exposed individuals. Blood levels of gamma-HCH were 0.002–0.34 mg/litre. The frequency of clinical symptoms and electroencephalographic changes was higher among individuals whose blood contained 0.02 mg/litre or more of gamma-HCH (Czeglédi-Janko & Avar, 1970; American Conference of Governmental Industrial Hygienists, Inc., 1986).

Tolot et al. (1969) and Schüttmann (1972) reported peripheral neuropathies after contact with technical-grade HCH or lindane.

3.2.2 Irritation and sensitization

Behrbohm & Brandt (1959) described 26 cases of allergic and toxic dermatitis in workers exposed during the manufacture of technical-grade HCH. Patch testing with pure alpha-, beta-, gamma-, and delta-HCH gave negative results, but positive reactions were obtained with residual fractions. Baumgartner (1953) also described skin sensitization in four workers involved in lindane manufacture. Cases of allergic disease (rhinitis, conjunctivitis, and eczema) were also reported among workers in the USSR exposed to lindane (Krzhyzhanovskaya, 1973, and Bezuglikh et al., 1976; see Izmerov, 1983).

Patch tests were performed with 1% lindane dissolved in a petroleum solvent on the upper back of 200 subjects, 105 men and 95 women, aged 18–76 years; 50 of them (34 men and 16 women) were agricultural workers,

Effects on humans

24 (18 men and six women) had worked on the land in the past, and the other 53 men and 73 women had never used pesticides. Results were read after 48 and 72 h. A positive reaction was found in none of the 200 subjects tested (Lisi et al., 1986).

In a study to establish the optimal test concentration of lindane and the frequency of irritant and sensitization reactions, Lisi et al. (1987) tested 335 men and women, of whom 70 were employed in agriculture, 25 had been employed in agriculture in the past and 240 had never been exposed to pesticides. The results of the patch test, using 1% lindane in a petroleum solvent, were all negative, and no sensitization reaction was observed.

9. EFFECTS ON OTHER ORGANISMS IN THE LABORATORY AND FIELD

9.1 Microorganisms

9.1.1 Bacteria

The effect of lindane at concentrations up to 100 times the recommended dose for field application (1 mg/kg of soil) was studied on organic mineralization and on nitrification rates in alluvial soil around Delhi, India. Lindane inhibited the evolution of CO_2 from soil at a concentration of 100 mg/kg but not or only slightly with 1 and 10 mg/kg over an incubation period of 120 days. Nitrification by the bacterial species *Nitrobacter* and *Nitrosomonas* was decreased with the dose of 100 mg/kg, and 10 mg/kg induced inhibition during the first three weeks after application. Nitrification was restored within 35 days (Gaur & Misra, 1977).

The influence of technical-grade HCH on the nitrifying characteristics of *Nitrosomonas* and *Nitrobacter* species isolated from an alluvial soil was studied in artificial growth media and in a flooded, autoclaved soil. The concentrations tested were 0, 5, 20, and 50 mg/kg soil or culture medium; the observation period was 10–40 days. The rate of nitrification was slower in autoclaved soil than in culture media. Technical-grade HCH inhibited nitrification at concentrations of 5 mg/kg and more (Ray, 1983).

Oxidation of substrate nitrogen was evaluated on the basis of the appearance of nitrite in the medium in the presence of *Nitrosomonas europaea* and on the disappearance of nitrite in the presence of *Nitrobacter agilis*. *N. agilis* was sensitive to lindane at concentrations as low as 1 µg/ml, and the highest concentration tested, 1000 µg/litre, delayed nitrate production. In *N. europaea*, lindane induced complete inhibition at a concentration of 10 µg/ml within 6–7 days (Garretson & San Clemente, 1968).

9.1.2 Algae

9.1.2.1 Blue-green algae

Bringmann & Kühn (1978) saw no inhibition of growth of *Microcystis aeruginosa* at concentrations of lindane up to 0.3 mg/litre, and Palmer & Maloney (1955) found no inhibition of the growth of *Microcystis*

Effects on other organisms in the laboratory and field

aeruginosa or *Cylindrospermum licheniforme* at 2 mg/litre. Both concentrations are higher than those that would result in adverse effects in fish or crustaceans.

Concentrations of 50, 60, and 80 µg/ml of lindane were lowered in the presence of 25 species of blue-green algae, thus reducing its toxicity (Das & Singh, 1977).

9.1.2.2 Freshwater algae

Lindane was lethal to *Scenedesmus acutus* (age of culture 1, 3, or 5 days) at a concentration of 5 mg/litre after 5 days' exposure. At concentrations of 1–5 mg/litre, 50% growth reduction occurred (Krishnakumari, 1977). Jeanne-Levain (1974) also observed about 50% inhibition of growth of *Dunaliella bioculata* with lindane at 5 mg/litre; 10 mg/litre completely inhibited growth. Similar findings were described by Jeanne (1979). No observable inhibition of growth was found by Bringmann & Kühn (1978) in *Scenedesmus quadricauda* at concentrations up to 1.9 mg/litre; and Palmer & Maloney (1955) saw no inhibition in *Scenedesmus obliquus*, *Chlorella variegata*, *Gomphonema parvulum*, or *Nitzschia palea* at the only dose tested, 2 mg/litre.

9.1.2.3 Marine algae

Growth inhibition was studied in three species of marine algae by Ukeles (1962), using concentrations of 1–9 mg/litre. No inhibitory effect was found in *Protococcus* sp. at up to the highest concentration tested or in *Pheodactylum tricornutum* at up to 1 mg/litre. *Chlorella* sp. were slight inhibited at the lowest concentration tested, but the dose–response relationship indicated that the threshold was not far below this dose. In tests with green algae and diatoms, growth was not inhibited by lindane at concentrations below about 1 mg/litre.

9.1.3 Dinoflagellates, flagellates, and ciliates

Representative members of these three groups were exposed to lindane at concentrations of 0.5–60 mg/litre (Jeanne-Levain, 1974). Lethality was reached in *Amphidinium carteri* (dinoflagellate) at 2 mg/litre and in *Tetrahymena pyriformis* (ciliate) at > 10 mg/litre; no lethal effect was observed in *Euglena gracilis* (flagellate) at up to the highest dose of 60 mg/litre. All doses had inhibitory effects.

Ukeles (1962) observed inhibition of the growth of the flagellates *Monochrysis lutheri* and *Dunaliella euchlora* with concentrations of lindane > 5 mg/litre.

9.2 Aquatic organisms

9.2.1 Invertebrates

Green et al. (1986) investigated a range of 10 common European freshwater macroinvertebrates in a continuous flow system to establish their response to lindane at concentrations ranging from 25 ± 4.8 to 430 ± 32 µg/litre under similar test conditions: water temperature, 11 ± 1 °C; pH, 7.5–8.0; dissolved oxygen, > 96%; and hardness of water (as $CaCO_3$), 92.9 ± 6.3 mg. The 96-h LC_{50} varied from 4.5 up to > 430 µg/litre (Table 14). The ephemeropteran *Baetis rhodani* and the plecopterans *Leuctra moselyi* and *Protonemura meyeri* were the most sensitive species; *Physa fontinalis* and *Polycelis tenuis* were the most tolerant.

Table 14. LC_{50} values for invertebrates

Taxon	Species	96-h LC_{50} (µg/litre)
Insecta		
Plecoptera	*Pteronarcys california*	4.5
	Leuctra moselyi	< 130
	Protonemura meyeri	< 130
Ephemeroptera	*Baetis rhodani*	54
Diptera	*Chironomus riparius*	235
Trichoptera	*Hudropsyche angustipennis*	330
Crustacea		
Amphipoda	*Gammerus rhodani*	48
	Gammerus facia	10
	Gammerus pulex	225
Isopoda	*Asellus aquaticus*	375
Mollusca		
Gastropoda	*Physa fontinalis*	> 430
Platyhelminthes		
Tricladia	*Polycelis tenuis*	> 430
Annelida		
Oligocheata	*Limnodrilus hoffmeisteri*	> 430

Effects on other organisms in the laboratory and field

9.2.1.1 Crustaceans

The sensitivity of both freshwater and seawater crustaceans to the acute toxicity of lindane is summarized in Table 15. In general, freshwater crustaceans were much less sensitive to lindane than those living in seawater. The LC_{50} values for freshwater crustaceans are comparable to or about one order of magnitude higher than the median values for fish, whereas seawater crustaceans are generally an order of magnitude more sensitive than fish.

Daphnia magna were exposed to lindane at 11–210 µg/litre continuously for 64 days, thus covering three succeeding generations (Macek et al., 1976). A clear dose-dependent depression of reproduction was found at higher concentrations. The authors assessed the NOEL to be between 11 and 19 µg/litre. In the same study, *Gammarus fasciatus* was exposed to lindane at concentrations of 1.2–17.7 µg/litre. The NOEL for survival and reproductive success was 4.3 µg/litre. *Gammarus fasciatus* is thus more sensitive to both the acute and chronic effects of lindane.

9.2.1.2 Aquatic arthropods

The results of studies on the acute toxicity of lindane in freshwater insects (mainly nymphs) are summarized in Table 15. Aquatic insects can be seen to be sensitive to lindane, with considerable differences between the tested species. The NOEL for continuous exposure of two successive generations of *Chironomus tentans* was 2.2 µg/litre (Macek et al., 1976).

The 48-h LC_{50} for the water mite *Hydrachna trilobata* Viets, an aquatic arachnida, was 0.05 mg/litre (Nair, 1981).

9.2.1.3 Molluscs

The acute effects of lindane have been investigated in bivalves and gastropods; no data were available on cephalopods. The LC_{50} values determined (Table 16) are all > 1 mg/litre, and therefore at least one order of magnitude higher than those reported for fish, indicating that molluscs are much less sensitive to the acute action of lindane.

Butler (1963b) determined shell growth in oysters exposed to various concentrations of lindane. When sufficiently irritated, oysters close their shells, do not feed and therefore do not grow. Even at the lowest dose tested, 1 mg/litre, a decreased shell growth of 43% was observed.

Table 15. Acute toxicity of gamma-HCH to freshwater and marine crustaceans

Species	Size (g, mm, cm, age)	Temperature (°C)	Test conditions	96-h LC$_{50}$ (µg/litre)	Reference
Freshwater					
Daphnia magna	24-h old	20	Static	516	Randall et al. (1979)
			Static	485	Juhnke & Lüdemann (1978)
	18-h old	20	Static	1100	Sanders & Cope (1966)
Daphnia pulex	18-h old	21	Static	460	Sanders & Cope (1966)
Gammarus fasciatus			Static	39	Juhnke & Lüdemann (1978)
Gammarus pulex	1.0–1.5 cm	15	Static[a]	34	Abel (1980)
Gammarus lacustris	2 months?	21	Static	48	Sanders (1969)
Simocephalus serrulatus	18-h old	21	Static	880	Sanders & Cope (1966)
Marine					
Pink shrimp (*Penaeus duorarum*)	30–52 mm	24–26	Flow-through	0.17	Schimmel et al. (1977)
Brown shrimp (*Penaeus aztecus*)	Adult	30	Flow-through	0.4	Butler (1963b)
Brown shrimp (*Crangon crangon*)	–	15	Static	1–3.3	Portmann (1970)
Grass shrimp (*Palaemonetes pugio*)		23–27	Flow-through	4.4	Schimmel et al. (1977)
Grass shrimp (*Palaemonetes vulgaris*)	0.47 g	20	Static	10	Eisler (1969)
Sand shrimp (*Crangon septemspinosa*)	0.25 g	20	Static	5	Eisler (1969)
Hermit crab (*Pagurus longicarpus*)	0.28 g	20	Static	5	Eisler (1969)
Mysid shrimp (*Mysidopsis bahia*)	8–9.5 mm	23–25	Flow-through	6.3	Schimmel et al. (1977)

[a] Water renewed regularly

Effects on other organisms in the laboratory and field

Table 16. Acute toxicity of gamma-HCH to aquatic invertebrates

Species	Freshwater/ Marine water	Size (g, mm, cm, age)	Temperature (°C)	Test conditions	96-h LC$_{50}$ (mg/litre)	Reference
Bivalves						
Mytulis galloprovincialis	Marine	50–70 mm	17	Static	5.5	Rao (1981)
Mercenaria mercenaria	Marine	Eggs	20	Static	> 10	Eisler (1970a)
	Marine	Eggs	24	Static	> 10[a]	Davis & Hidu (1969)
Crassostrea virginica	Marine	Eggs	24	Static	9.1[a]	Davis & Hidu (1969)
Cardium edule	Marine	–	15	Static	10[a]	Portmann (1970)
Gastropods						
Physa acuta	Freshwater	–	20–28	Static	8.1[a]	Hashimoto & Nishiuchi (1981)
Semisulcospira libertina	Freshwater	–	20–28	Static	6.2[a]	Hashimoto & Nishiuchi (1981)
Indoplanorbis exastus	Freshwater	–	20–28	Static	7.1[a]	Hashimoto & Nishiuchi (1981)
Lymnea stagnalis	Freshwater			Static[b]	7.3[a]	Bluzat & Seuge (1979)
Insects						
Cloeon dipterum	Freshwater	–	20–28	Static	0.15[a]	Hashimoto & Nishiuchi (1981)
Chironomus tentans	Freshwater			Static	0.207[a]	Juhnke & Lüdemann (1978)
Pteronarcys californica	Freshwater	30–35 mm	15.5	Static	0.0045	Sanders & Cope (1968)

[a] 48-h LC$_{50}$
[b] Water renewed regularly

The freshwater snail *Lymnea stagnalis* was exposed to lindane at concentrations of 1 or 2 mg/litre for periods up to seven weeks, and therefore at least two successive generations. No increase in mortality over that in controls was observed at either concentration. A slight decrease in shell growth was found at 2 mg/litre, and egg production was reduced at 1 and 2 mg/litre. Embryonic development was also disturbed at both concentrations in a dose-dependent manner. A NOEL for reproduction was not obtained in this study, but the results suggest that it is close to 1 mg/litre (Bluzat & Seuge, 1979; Seuge & Bluzat, 1979a,b, 1982).

In a study on the effects of lindane on egg production in the mud snail *Nassa obsoleta* (Eisler, 1970a), the number of egg cases deposited by day 33 after treatment following initial exposure for 96 h was measured. A clear NOEL was found at 1 mg/litre; the next dose tested (10 mg/litre) induced a clear decrease.

Davis & Hidu (1969) found 67% survival of the larvae of the hard clam *Mercenaria mercenaria* over 12 days after exposure to lindane at a concentration of 10 mg/litre. For oysters (*Crassostrea virginica*), 50% survival was seen after exposure at 9.1 mg/litre over 48 h.

Overall, these studies show that reproduction of molluscs is not adversely affected at concentrations just below 1 mg/litre, a level much higher than the NOELs for fish and crustaceans.

1.2.2 Fish

1.2.2.1 Acute toxicity

The results of studies on the acute toxicity of lindane in fish have been reported in the literature since 1959, although in most of these no data were given on the purity of the lindane used. Since there is no significant difference in the values reported around 1960 and those reported in 1975–80, however, data on the acute toxicity of lindane are summarized here regardless of whether the purity of the compound tested was reported.

LC_{50} values for lindane in several fish species are summarized in Table 17. Most values fall within a range of 0.02–0.09 mg/litre, the majority being around 0.05 mg/litre. Only Macek & McAllister (1970) found an extraordinarily low LC_{50} value for brown trout of 0.002 mg/litre.

The symptoms of acute poisoning are mainly gross irritability, loss of equilibrium, changes in pigmentation and localized peripheral haemorrhage.

Effects on other organisms in the laboratory and field

Irritability appears within the first minutes of exposure and is accompanied by loss of equilibrium and disturbed swimming motion. Poisoned fish show signs of respiratory distress. Haemorrhages appear at sub-lethal doses 2–4 days after the beginning of exposure.

A clear temperature dependence of the LC_{50} was found in bluegills (Cope, 1965; Macek et al., 1969), lindane being more toxic at higher temperatures, although the range of LC_{50} values was the same. Macek et al. (1969) found LC_{50} values of 0.054 mg/litre at 12.7 °C and 0.037 mg/litre at 23.8 °C.

In two studies, wild populations of mosquito fish were compared to laboratory strains and to each other with respect to the toxic action of lindane (Culley & Ferguson, 1969). Wild populations from areas that had previously been treated with lindane tolerated higher concentrations of the substance: one wild population had a LC_{50} value of 3.104 mg/litre, whereas a susceptible population from a different region had a value of 0.074 mg/litre. These results suggest that wild populations can adapt to lindane when exposed repeatedly. Boyd & Ferguson (1964) found a similar effect.

9.2.2.2 Long-term toxicity

Macek et al. (1969) exposed bluegills to lindane at 0.6–9.1 µg/litre for 18 months. No adverse effect was observed at any of the tested concentrations; the NOEL was therefore considered to be > 9.1 µg/litre. In the same study, fathead minnows (*Pimephales promelas*) were exposed to lindane at 1.4–23.5 µg/litre for 43 weeks. A statistically significant increase in mortality was observed at the highest dose. Growth of the surviving fish was not adversely affected, and spawning appeared to be normal in all test groups. The NOEL in this experiment was considered to be 9.1 µg/litre.

Macek et al. (1969) exposed brook trout (*Salvelinus fontinalis*) to lindane at concentrations of 1.0–16.6 µg/litre for 261 days. Only slight effects on growth were observed at the end of the exposure period. Although no statistical assessment of spawning was performed, fish exposed to the highest test level were adversely affected in this respect, and the NOEL was set at 8.8 µg/litre.

The results of these studies indicate that concentrations not far below those inducing 50% mortality are well tolerated for long periods (up to 18 months). A 5–10-fold difference can be seen between the maximum

Table 17. Acute toxicity of gamma-HCH to fish

Species	Freshwater/ Marine water	Size (g, mm, cm)	Temperature (°C)	Test conditions[a]	96-h LC$_{50}$ (µg/litre)	Reference
Rainbow trout (*Salmo gairdneri*)	Freshwater	0.69 g	12	Static	32	McLeay (1976)
	Freshwater	3.2 g	20	Static	38	Katz (1961)
		0.6–1.7 g	13	Static	27	Macek & McAllister (1970)
		0.7 g	13	Static	22	Cope (1965)
		3-cm fry	12	Flow-through	22	Tooby & Durbin (1975)
		Yearling	12	Flow-through	30	Tooby & Durbin (1975)
		2.6 g	15	No detail	34[b]	Dion (1984)
Brown trout (*Salmo trutta*)	Freshwater	0.6–1.7 g	13	Static	2	Macek & McAllister (1970)
		1.1 g	15	No detail	38[b]	Dion (1984)
Coho salmon (*Oncorhynchus kisutch*)	Freshwater	2.7–4.1 g	20	Static	50	Katz (1961)
		0.6–1.7 g	13	Static	41	Macek & McAllister (1970)
Chinook salmon (*Oncorhynchus tschawytscha*)	Freshwater	1.5–5 g	20	Static	40	Katz (1961)
Bluegill (*Lepomis macrochirus*)	Freshwater	0.6–1.7 g	18	Static	68	Macek & McAllister (1970)
		1.0 g	18	Static	53	Cope (1965)
		0.26 g	19	Static	57	Randall et al. (1979)
		–	25	Static	77	Henderson et al. (1959)
		0.6–1.5 g	18	Static	51	Macek et al. (1969)
Redear sunfish (*Lepomis microlophus*)	Freshwater	0.6–1.7 g	18	Static	83	Macek & McAllister (1970)
Threespine stickleback (*Gasterosteus aculeatus*)	Freshwater	0.38–0.77 g	Room temperature	Static	44 and 50[c]	Katz (1961)
Carp (*Cyprimus carpio*)	Freshwater	0.6–1.7 g	18	Static	90	Macek & McAllister (1970)
		6.8 cm	17–19	Static	280[b]	Lüdemann & Neumann (1960a)
		–	–	Static	310[b]	Hashimoto & Nishiuchi (1981)

Effects on other organisms in the laboratory and field

Table 17 (contd).

Species	Freshwater/ Marine water	Size (g, mm, cm)	Temperature (°C)	Test conditions	96-h LC$_{50}$ (µg/litre)	Reference
Guppy (Lebistes reticulatus)	Freshwater	–	25	Static	138	Henderson et al. (1959)
		0.37 g (male)	24	No detail	160	Boulekbache (1980)
		0.5 g (female)	24	No detail	300[b]	Boulekbache (1980)
Golden orfe (Leuciscus idus melanotus)	Freshwater	No detail	No detail	Static	30 and 280[b]	Juhnke & Lüdemann (1978)
Goldfish (Carassius auratus)	Freshwater	0.6–1.7 g	18	Static	131	Macek & McAllister (1970)
		–	25	Static	152	Henderson et al. (1959)
		–	–	Static	120[b]	Hashimoto & Nishiuchi (1981)
Fathead minnow (Pimephales promelas)	Hard freshwater	0.6–1.7 g	18	Static	87	Macek & McAllister (1970)
	Soft freshwater	–	25	Static	62	Henderson et al. (1959)
Largemouth bass (Micropterus salmoides)	Freshwater	0.6–1.7 g	18	Static	32	Macek & McAllister (1970)
Channel catfish (Ictalurus punctatus)	Freshwater	0.6–1.7 g	18	Static	44	Macek & McAllister (1970)
Back bullhead (Ictalurus melas)	Freshwater	0.6–1.7 g	18	Static	64	Macek & McAllister (1970)
Yellow perch (Perca flavescens)	Freshwater	0.6–1.7 g	18	Static	68	Macek & McAllister (1970)
Atlantic silverside (Menidia menidia)	Marine water	2 g	20	Static	9[d]	Eisler (1970b)
Bluehead (Thalassoma bifasciatum)	Marine water	7 g	20	Static	14[d]	Eisler (1970b)
Striped killifish (Fundulus majalis)	Marine water	1.6 g	20	Static	28[d]	Eisler (1970b)

Table 17 (contd).

Species	Freshwater/ Marine water	Size (g, mm, cm)	Temperature (°C)	Test conditions	96-h LC$_{50}$ (µg/litre)	Reference
Longnose killifish (*Fundulus similis*)	Marine water	–	29	Static	240[b]	Butler (1963b)
Mummichog (*Fundulus heteroclitus*)	Marine water	2 g 42 mm (approx. 0.8 g)	20 20	Static Static	60[d] 20[d]	Eisler (1970b) Eisler (1970c)
Striped mullet (*Mugil cephalus*)	Marine water	6.6 g	20	Static	66[d]	Eisler (1970b)
White mullet (*Mugil curena*)	Marine water	–	16	Static	30[b]	Butler (1963b)
American eel (*Anguilla rostrata*)	Marine water	0.18 g	20	Static	56[d]	Eisler (1970b)
Eel (*Anguilla anguilla*)	Freshwater	50–80 g	11.5	Static	600[b]	Foulquier et al. (1971)
Northern puffer (*Sphaeroides maculatus*)	Marine water	100 g	20	Static	35[d]	Eisler (1970b)
Striped bass (*Morone saxatilis*)	Freshwater	0.93 g 2.4 g	21 13	Static Static	400 7.3	Wellborn (1971) Korn & Earnest (1974)
Harlequin fish (*Rasbora heteromorpha*)	Freshwater	1.3–3.0 cm	20	Static	45[b]	Alabaster (1969)
Mosquito fish (*Gambusia affinis*)	Freshwater	– –	21 ca22	Static Static	150[a] 74[b]	Boyd & Ferguson (1964) Culley & Ferguson (1969)
Sheepshead minnow (*Cyprinodon variegatus*)	Marine water	17–21 mm	24–28	Flow-through	104	Schimmel et al. (1977)
Pinfish (*Lagodon rhomboides*)	Marine water	42–61 mm	22.5–25	Flow-through	30.6	Schimmel et al. (1977)

Table 17 (contd).

Species	Freshwater/ Marine water	Size (g, mm, cm)	Temperature (°C)	Test conditions	96-h LC$_{50}$ (µg/litre)	Reference
Spot (*Leiostomus xanthurus*)	Marine water	—	15	Flow-through	30[b]	Butler (1963a)
Japanese rice fish (*Oryzias latipes*)	Freshwater	—	—	Static	120[b]	Hashimoto & Nishiuchi (1981)
Pond loach (*Misgurnus anguilicaudatus*)	Freshwater	—	—	Static	150[b]	Hashimoto & Nishiuchi (1981)
Gudgeon (*Gobio gobio*)	Freshwater	—	12.2	Flow-through	73	Marcelle & Thorne (1983)
Viviparus bengalensis	Freshwater	2.5–3.4 g	24	No detail	1486	Panwar et al. (1982)

[a] Water renewed regularly
[b] 48-h LC$_{50}$
[c] Depending on salinity
[d] 24% salinity, pH 8.0
[e] 36-h LC$_{50}$

dose tolerated in long-term tests and the LC$_{50}$ in the three species tested (Macek et al., 1976).

9.2.2.3 Reproduction

The reproductive effects of lindane were tested in bluegills, fathead minnows, and brook trout (Macek et al., 1969). Spawning, hatchability of eggs, and survival of the fry appeared not to be adversely affected by concentrations of up to 9.1 µg/litre in guppies, up to 23.4 µg/litre in fathead minnows, and up to 2.1 µg/litre in brook trout.

9.2.3 Amphibia

9.2.3.1 Acute toxicity

All of the published studies on the acute toxicity of lindane in amphibia (Table 18) were undertaken with tadpoles. Tests on tadpoles are considered to be the most reliable for assessing possible adverse effects to aquatic organisms since these larvae live exclusively in an aquatic environment, whereas adults spend a major part of their lifetime outside the water. The results show that larvae of amphibia are less sensitive to lindane than fish.

9.2.3.2 Effects on hatching and larval development

Marchal-Segault & Ramade (1981) exposed eggs and larvae of *Xenopus laevis* to lindane at concentrations of 0.5–2 mg/litre in tap water. Hatchability was reduced by the highest dose only; however, development of the larvae was disturbed at all concentrations testing, as seen by lowered body weights, longer periods (four weeks) from hatching to metamorphosis, and morphological abnormalities. Altered function of the hypothalamo-hypophyseal axis, which regulates growth and metamorphosis, and dysfunction of the intermediate lobe of the hypophysis, which controls pigmentation, were suggested. A NOEL could not be established.

Effects on other organisms in the laboratory and field

Table 18. Acute toxicity (48-h LC_{50}) of lindane in tadpodes of freshwater amphibia under static conditions

Species	Size (g, mm, cm, age)	Temperature (°C)	LC_{50} (mg/litre)	Reference
Pseudacris triserata	7-days old	15.5	3.8	Sanders (1970)
Bufo woodhousii	4–5-weeks old	15.5	5.4	Sanders (1970)
Bufo bufo japonicus	—	—	24	Hashimoto & Nishiuchi (1981)
Bufo bufo L.	25–30 mm	18–21	0.3	Lüdemann & Neumann (1960b)

9.3 Terrestrial organisms

9.3.1 Honey-bees

Atkins et al. (1973) estimated the LD_{50} in the honey-bee to be 0.56 µg.

9.3.2 Birds

9.3.2.1 Acute toxicity

Some of the studies of the acute toxicity of lindane in birds were undertaken soon after its introduction as an insecticide, and, in these, the quality of lindane used is not specified. Nevertheless, these studies are included in this review, as well as studies in which no precise LD_{50} could be obtained.

The results are summarized in Table 19. The values obtained cover a wide range, but most are in the order of 100 mg/kg body weight. Common symptoms of poisoning are vomiting and loss of appetite, loss of weight, and hyperexcitability; central nervous system symptoms occur as incoordination, convulsions, and tremors (Rosenberg et al.,1953; Dahlen & Haugen, 1954; Adamic, 1958; Turtle et al., 1963; Dittrich, 1966).

In hens, lethal oral doses of 330–1440 mg/kg body weight caused inflammation of the gastrointestinal tract, degeneration of the liver and kidneys, and changes in ganglionic cells (Adamic, 1958). Similar findings were obtained in bobwhite quails and mourning doves (Rosenberg et al., 1953; Dahlen & Haugen, 1954) at 120–210 mg/kg body weight. In doves, doses of > 300 mg/kg body weight caused mainly liver atrophy, congested lungs and kidneys, and haemorrhage.

The concentrations of lindane in the diet that caused 50% mortality in young and adult bobwhite quail, ring-necked pheasants and mallard ducks within < 10 days and < 100 days (Dewitt et al., 1963) are summarized in Table 20.

9.3.2.2 Short-term toxicity

Chen & Liang (1956) fed white leghorn chickens and hybrid native ducks diets containing lindane at 2, 4, or 10 mg/kg of diet for three months. No adverse effect was observed in either species at any dose.

Effects on other organisms in the laboratory and field

Table 19. Toxicity of lindane to birds

Species	Parameter	Concentration (mg/kg)[a]	Reference
Bobwhite quail (Colinus virginianus)	5-d LC$_{50}$ acute LD$_{50}$ acute LD$_{50}$	882 (755–1041) 120–130 (male) 190–210 (female)	Hill et al. (1975) Dahlen & Haugen (1954) Dahlen & Haugen (1954)
Japanese quail (Coturnix coturnix japonica)	5-d LC$_{50}$	490 (408–589) 205 and 425 (347–520)	Hill & Camardese (1986) Clausing et al. (1980)
Ring-necked pheasant (Phasianus colchicus)	5-d LC$_{50}$	561 (445–690)	Hill et al. (1975)
Mallard (Anas platyrhynchos)	acute LD$_{50}$ 5-d LC$_{50}$	> 2000 (male) > 5000	Hudson et al. (1984) Hill et al. (1975)
Starling (Sturnus vulgaris)	acute LD$_{50}$	100	Schafer (1972)
Red-winged blackbird (Agelaius phoeniceus)	acute LD$_{50}$	75	Schafer (1972)
Common grackle (Quiscalus quiscula)	acute LD$_{50}$	> 100	Schafer (1972)
House sparrow (Passer domesticus)	acute LD$_{50}$	56 (320–100)	Schafer (1972)
Common crow (Corvus brachyrhynchos)	acute LD$_{50}$	> 100	Schafer (1972)
Mourning dove (Streptopelia risoria)	acute LD$_{50}$	350–400	Dahlen & Haugen (1954)
Feral pigeon (Columba livia)	acute LD$_{50}$	> 600	Turtle et al. (1963)

[a] Acute LD$_{50}$ expressed as milligrams per kilogram body weight in a single oral dose; otherwise, concentration expressed as milligrams per kilogram food (i.e., birds were fed with a dosed diet for 5 days followed by a "clean" diet for 3 days)

Table 20. Oral LD$_{50}$ of lindane in birds[a]

Species	Lindane intake (mg/kg body weight)	
	< 10 days	< 100 days
Young bobwhite	1070	930
Adult bobwhite	–	1050
Young ring-necked pheasant	175	> 1800
Adult ring-necked pheasant	–	> 630
Young mallard	415	–
Adult mallard	1000	–

[a] From Dewitt et al. (1963)

When laying hens were fed diets containing lindane at 0.01, 0.1, 1, or 10 mg/kg of diet for 60 days, no effect was observed on body weight gain, mortality, clinical symptoms, or egg production. The authors concluded that lindane does not adversely affect reproduction in hens at doses up to 10 mg/kg of diet (Ware & Naber, 1961).

Harrison et al. (1963) fed diets containing lindane at 4, 16, or 64 mg/kg to white Leghorn x Australorp chickens for 27 days. Increased mortality was seen in the highest dose group, and the two higher doses resulted in enlarged livers. No pathological change was found in the animals given the lowest dose, but in the two higher dose groups dose-dependent liver hypertrophy was observed. Tissues were not examined microscopically. The NOEL in this experiment was concluded to be 4 mg/kg of diet.

The 30-day oral LD$_{50}$ for male mallard ducks (12 animals) was 30 mg/kg body weight; as the acute LD$_{50}$ was > 2000 mg/kg body weight, the toxic action of lindane appears to be cumulative (Hudson et al., 1984).

The LC$_{50}$ values for lindane given in the diet for five days were 882 mg/kg of diet in bobwhite quail (aged 9 days), 425 mg/kg of diet in Japanese quail (aged 7 days), 561 mg/kg of diet in ring-necked pheasant (aged 10 days), and > 5000 mg/kg of diet in mallard ducks (aged 15 days) (Hill et al., 1975).

(a) Effect on egg-shell quality: Whitehead et al. (1972a,b, 1974) found that the shells of hens' eggs were not adversely affected by administration of lindane in amounts up to 200 mg/kg of diet; however, egg production was reduced at 100 and 200 mg/kg of diet. Doses of up to 100 mg/kg of diet had no effect on hatchability, egg weight, yolk weight, shell thickness, calcium content, shearing strength or structure. The NOEL was 10 mg/kg of diet. Similar findings were obtained in Japanese quail.

Effects on other organisms in the laboratory and field

(b) Field experience: A population of Canada geese (*Branta canadensis*) living in the Pacific Northwest of the USA was observed from 1978 through 1981. Lowered reproductive success, increased mortality among adults and a population decline in this region were associated with the use of heptachlor for treating wheat seed. This hypothesis was supported by the results of analyses of eggs and tissues. In 1979, heptachlor was replaced by lindane for use in this area; the reproductive success of the geese increased, mortality decreased, and the population increased. There was no evidence of either biomagnification of lindane from treated seed to goose tissues or eggs or of induction of adverse effects by lindane (Blus et al., 1984).

9.3.2.3 Reproduction

Lindane (99.8% in olive oil) was administered by stomach tube to four groups of laying ducks (*Anas platyrhynchus domesticus*), comprising one drake and four ducks, at doses of 0 or 20 mg/kg body weight daily, three times per week, or twice a week for eight weeks. Egg laying stopped immediately in the groups treated daily and three times weekly and was irregular when it resumed, with drastically reduced clutch sizes. The effect in the group treated twice weekly was marginal. At the end of treatment, the laying frequencies for the four groups were 50%, 8.3%, 11.7%, and 40%, respectively. Hepatic, plasma, and ovarian vitellogenin levels were reduced in the groups treated daily and three times per week; the ovaries of the birds in these groups lacked mature vitellogenic follicles, and the thecal layer of moderately differentiated oocytes became highly atrophic. Levels of liver RNA were markedly reduced. A single injection of stilboestrol at 50 mg/kg body weight restored egg laying and the other parameters to normal within 24 h, suggesting that lindane imposed its effects by inducing oestradiol insufficiency (Chakravarty et al., 1986).

9.3.3 Mammals

The toxicity of lindane to bats has been studied because of its use in timber treatment. Racey & Swift (1986) exposed pipistrelle bats (*Pipistrellus pipistrellus*) to 1% lindane, both in combination with 5% pentachlorophenol in an organic solvent and alone. When it was applied in combination with pentachlorophenol, at the rates recommended by the manufacturer, to wooden roosting boxes six weeks before bats used them, the animals were killed within seven days; when the combination was applied 14 months before use, the animals died within 23 days. When lindane was applied alone two weeks before exposure of bats in the boxes, all animals died within four days. These results were significant at the 0.1% level.

Boyd et al. (1988) exposed pipistrelle bats to wood blocks coated with lindane at 9.9 mg/m^2 for 44 days and then for a further 44 days to blocks coated with lindane at 866 mg/m^2, 24 h after coating the blocks. Significant mortality ($p < 0.007$) was recorded. In a second experiment, all bats exposed to lindane at either 147 or 211 mg/m^2 died within 17 days, whereas no death occurred among controls exposed to the solvent only.

Turner (1979) studied the distribution and concentration of gamma-HCH in maternal and fetal tissues of a 6.5-year-old desert bighorn (*Ovis canadensis cremnobates*). The maternal organs and tissues and the tissues of the term ram fetus contained residue levels ranging from none detected to 0.01 mg/kg on a fat basis. Residues of 0.01 mg/kg were present in adipose tissues, muscle, liver, gonads, and placenta. Placental transfer of gamma-HCH thus appears to be very low.

9.4 Appraisal

The toxicity of lindane to organisms in the environment must be assessed on the basis of the results of laboratory toxicity tests and of the probable bioavailability of lindane to similar organisms exposed in the wild. The strong adsorption of lindane to particles might be expected to reduce its toxic effects below that seen in laboratory studies of microorganisms in culture and of aquatic organisms in water without sediment; however, there is insufficient information to substantiate this hypothesis. No information was available on the toxicity of lindane to organisms that feed on or live in sediment.

Low levels of residues in birds in the wild, coupled with the reported low toxicity of lindane in laboratory tests, make it unlikely that it affects birds in the wild.

Bats are killed by applications of lindane to wood at normal rates and are affected by residues of past wood treatment. Since many bat species are declining in numbers or are extremely rare, lindane must be regarded as a major hazard and its use avoided in areas where bats might be found. Other mammals are unlikely to be adversely affected by this compound.

10. PREVIOUS EVALUATIONS BY INTERNATIONAL BODIES

The International Agency for Research on Cancer (1987) evaluated the hexachlorocyclohexanes and concluded that there is sufficient evidence for the carcinogenicity to experimental animals of the technical grade and the alpha isomer; such evidence was considered to be limited for the beta and gamma isomers. There is considered to be inadequate evidence for their carcinogenicity in humans. Hexachlorocyclohexanes were thus classified into group 2B, possibly carcinogenic to humans.

WHO (1990) classified technical-grade lindane as 'moderately hazardous' in normal use, on the basis of an LD_{50} of 88 mg/kg. WHO/FAO (1975) issued a data sheet on lindane (No. 12), dealing with labelling, safe handling, transport, storage, disposal, decontamination, training, and medical supervision of workers, first-aid and medical treatment.

Lindane was evaluated by the FAO/WHO Joint Meeting on Pesticide Residues in 1966, 1967, 1968, 1969, 1973, 1974, 1975, 1977, 1979, and 1989 (FAO/WHO, 1967, 1968, 1969, 1970, 1974, 1975, 1976, 1978, 1980, 1990). A maximal acceptable daily intake of lindane in humans was established at 0–0.008 mg/kg body weight by the 1989 Joint Meeting (WHO, 1990). This value is based on a NOAEL of 10 mg/kg in the diet, equivalent to 0.75 mg/kg body weight per day in rats and 1.6 mg/kg body weight per day in dogs. Maximum residue limits have been recommended for more than 35 commodities, ranging from 0.01 mg/kg in milk to 3 mg/kg on strawberries; a limit of 0.5 mg/kg was recommended for most fruit and vegetables (Codex Alimentarius Commission, 1986; Table 21).

Table 21. Maximum residue limits (MRL) for gamma-HCH of the Codex Alimentarius Commission (1986)

Crop/Commodity	MRL (mg/kg)
Apples	0.5
Beans (dried)	1
Brussels sprouts	0.5
Cabbage	0.5
Cabbage, Savoy	0.5
Carrots	0.2 E[b]
Cattle, carcase meat (in the carcase fat)	2
Cauliflower	0.5
Cereal grains (including rice)	0.5
Cherries	0.5
Cocoa beans	1
Cocoa butter	1
Cocoa mass	1
Cranberries	3
Currants (red)	0.5
Eggs (on a shell-free basis)	0.1 E
Endive	2
Grapes	0.5
Kohlrabi	1
Lettuce	2
Milk	0.01
Pears	0.5
Peas	0.1
Pigs, carcase meat (in the carcase fat)	2
Plums	0.5
Potatoes	0.05[a]
Poultry (in the carcase fat)	0.7 E
Radishes	1
Rapeseed	0.05[a]
Sheep, carcase meat (in the carcase fat)	2
Strawberries	3
Sugar beets (roots)	0.1
Sugar beets (tops)	0.1
Spinach	2
Tomatoes	2

[a] Level at or about the limit of determination
[b] E, Extraneous residue limit

APPENDIX I. CHEMICAL STRUCTURE

The basic structure of HCH is a closed chain of six carbon atoms. The structure can have two spatial forms, a *cis* and a *trans* configuration. Each carbon atom is bound to a hydrogen and a chlorine atom, and one of these substituents forms a plane with the two connecting carbon atoms. Since this plane is parallel to the 'equator' of the molecule, this atom is said to be in the equatorial position. The bond with the other atom is parallel to the 'axis' of the molecule and is thus in the axial position. Owing to the size of the chlorine atom, the carbon atoms are not free to rotate, so the positions of the chlorine and hydrogen atoms are fixed: one is always in the equatorial position and the other in the axial position.

The various combinations of the spatial orientations of the hydrogen and chlorine atoms on each of the carbon atoms of cyclohexane result in different isomeric compounds. Theoretically, 17 isomers of HCH are possible; but, owing to spatial incompatibilities and thermodynamic instability, only nine isomers have been detected. They all have the *trans* configuration. In the beta isomer, all of the chlorine atoms are in the equatorial position. The positions in the major isomers of HCH are shown in Table 22 (Demozay & Marechal, 1972; Van Velsen, 1986).

Table 22. Positions of chlorine atoms in the major isomers of HCH[a]

Isomer	Chlorine position[b]	Physical structure
α[c]	AAEEEE	Monoclinic prisms
β	EEEEEE	Octahedral cubic
λ	AAAEEE	Monoclinic crystals
δ	AEEEEE	Crystals/fine patelets
ε	AEEAEE	Monoclinic needles or hexagonal monoclinic crystals

[a]From van Velsen (1986)
[b]A, axial position; E, equatorial position
[c]Racemate of two optical isomers

REFERENCES

Abalis IM, Eldefrawi ME, & Eldefrawi AT (1985) High-affinity stereospecific binding of cyclodiene insecticides and γ-hexachlorocyclohexane to γ-aminobutyric acid receptors of rat brain. Pestic Biochem Physiol 24: 95–102.

Abalis IM, Eldefrawi ME, & Eldefrawi AT (1986) Effects of insecticides on GABA-induced chloride influx into rat brain microsacs. J Toxicol Environ Health 18: 13–23.

Abel PD (1980) Toxicity of γ-hexachlorocyclohexane (lindane) to *Gammarus pulex*: Mortality in relation to concentration and duration of exposure. Freshwater Biol 10: 251–259.

Adamic S (1958) [The influence of gamma isomers of hexachlorocyclohexane on hen and duck.] Veterinarija (Sarajevo) 7: 329–334 (in Yugoslav).

Adams M, Coon FB, & Poling CE (1974) Insecticides in the tissues of four generations of rats fed different dietary fats containing a mixture of chlorinated hydrocarbon insecticides. J Agric Food Chem 22(1): 69–75.

Advisory Committee on Pesticides and Other Toxic Chemicals (1969) Further review of certain persistent organochlorine pesticides used in Great Britain. London, Her Majesty's Stationery Office.

Agnihotri NP, Pandey SY, Jain HK, & Srivastava KP (1977) Persistence of aldrin, dieldrin, lindane, heptachlor, and pp'-DDT in soil. J Entomol Res 1(1): 89–91.

Ahmed FE, Hart RW, & Lewis NJ (1977) Pesticide induced DNA damage and its repair in cultured human cells. Mutat Res 42: 161–174.

Aiyar AS (1980) Biological implications of pesticides: studies with lindane. Manage Environ pp. 182–192.

Alabaster JS (1969) Survival of fish in 164 herbicides, insecticides, fungicides, wetting agents, and miscellaneous substances. Int Pest Control 11(2): 29–35.

Albahary C, Dubrisay J, & Guerin (1957) Pancytopenic rebelle au lindane [isomère γ de l'hexachlorocyclohexane]. Arch Mal Prof Méd Trav Sécur Soc 18: 687–691.

Al-Omar MA, Al-Bassomy M, Al-Ogaily NV, & Al-Din Shebl D (1985) Residue levels of organochlorine insecticides in lamb and beef from Bagdad. Bull Environ Contam Toxicol 34: 509–512.

Al-Omar MA, Abdul-Jalil FH, Al-Ogaily NH, Tawfiq SJ, & Al-Bassomy MA (1986) A follow-up study of maternal milk contamination with organochlorine insecticide residues. Environ Pollut A42: 790–791.

American Conference of Governmental Industrial Hygienists Inc. (1986) Documentation of the threshold limit values and biological exposure indices, 5th ed. Cincinnati, Ohio, p. 348.

Anderson D & Styles JA (1978) The bacterial mutation test. Br J Cancer 37: 924–930.

Angerer J & Barchet R (1983) [α-, β- and γ-HCH in serum.] In: Henschler D, ed. [Analytical methods for the testing of working media prejudicial to health. Analyses in biological material, 7th instalment.] Weinheim, Verlag Chemie, Vol. 2 (in German).

Angerer J, Maasz R, & Heinrich R (1983) Occupational exposure to hexachlorocyclohexane. VI. Metabolism of γ-hexachlorocyclohexane in man. Int Arch Occup Environ Health 52: 59–67.

References

Anon (1984) [Organochlorine compounds including PCB: 1984 nitrition report.] Frankfurt am Main, German Nutrition Association (in German).

Anon (1989) FDA industry trying to cut pesticide contamination of lanolin. Pest Toxic Chem News, 3 May, pp. 26–27.

Antonovic EA (1958) Evaluation of the toxicity of gamma-isomer of hexachlorocyclohexane and its standardization in foodstuffs. Vopr Pitan **17**(6): 54–59.

Arbeitsgemeinschaft für die Reinhaltung der Elbe (1982) [Chlorinated hydrocarbons; data for the River Elbe. Report on the result of the programme for the measurement of chlorinated hydrocarbons in the section of the Elbe between Schnackenburg and the North Sea, 1980–1982]. Hamburg, Study Group for Keeping the Elbe Clean, pp. 64–65, 84, 86–94 (in German).

Arbeitsgemeinschaft für die Reinhaltung der Elbe (1988) [Water quality data for the Elbe from Schnackenburg to the sea. Numerical table, 1988.] Hamburg, Study Group for Keeping the Elbe Clean, p. 158 (in German).

Artigas F, Martinez E, Camon L, Gelpi E, & Rodriguez-Farre E (1988) Brain metabolites of lindane and related isomers: Identification by negative ion mass spectrometry. Toxicology **49**: 57–63.

Ashwood-Smith MJ, Trevino J, & Ring R (1972) Mutagenicity of dichlorvos. Nature **240**: 418–419.

van Asperen K (1958) [Distribution and metabolism of hexachlorocyclohexane in mammals.] In: [Proceedings of the IVth International Congress on Crop Protection, Hamburg, 8–15 September 1957.] Braunschweig, ACO Druck GmbH, Vol. 2, pp. 1619–1623 (in German).

Association of Official Government Chemists (1980) AOAC Methods: 29. Pesticide residues. Multiresidue methods: general method for organochlorine and organophosphorus pesticides (1), Washington, DC, pp. 466–475.

Atkins EL, Greywood EA, & MacDonald RL (1973) Toxicity of pesticides and other agricultural chemicals to honey bees. Riverside, California, University of California, Agricultural Extension (Unpublished report M-16).

Atlas E & Giam CS (1981) Global transport of organic pollutants: Ambient concentrations in the remote marine atmosphere. Science **211**: 163–165.

Bacci E, Calamari D, Gaggi C, Fanelli R, Focardi S, & Morosini M (1986) Chlorinated hydrocarbons in lichen and moss samples from the Antarctic Peninsula. Chemosphere **15**(6): 747–754.

Balaschow WE (1964) [Haematopoietic disorders following the action of the γ-isomer of hexachlorocyclohexane.] Gig Tr Prof Zabol **8**: 55–56 (in Russian).

Balba MH & Saha JG (1974) Metabolism of lindane-^{14}C by wheat plants grown from treated seed. Environ Lett **7**(3): 181–194.

Baluja G, Murado MA, & Tejedor MC (1975) Adsorption–desorption of lindane and aldrin by soils as affected by soil main components. Environ Qual Saf Suppl. 3: 243–249.

Barke A (1950) [Toxicity of the gamma-isomers of hexachlorocyclohexane.] Tierärztl Umsch **5**: 61–63 (in German).

Baron RL, Copeland F, & Walton MS (1975) Pharmacokinetics of organochlorine pesticides in mammalian adipose tissue. Environ Qual Saf Suppl. 3: 855–860.

Bauer U (1972) [Cncentration of insecticides, chlorinated hydrocarbons and PCB in algae. Schr Reihe Verh Wasser- Boden-Lufthyg **37**: 211–219 (in German).

Baumann K, Behling, K, Brassow HL, & Stapel K (1981) Occupational exposure to hexachlorocyclohexane. III. Neurophysiological findings and neuromuscular function in chronically exposed workers. Int Arch Occup Environ Health **48**: 165–172.

Baumgartner O (1953) [Sensitization to hexachlorocyclohexane.] Schweiz Med Wochenschr **83**(45): 1093–1094 (in German).

Bednarek W, Hansdorf W, Jörissen V, Schulte E, & Wegener H (1975) [The effects of chemical pollution of the environment on birds of prey in two test areas of Westphalia.] J Ornitol **116**(2): 181–194 (in German).

Behrbohm P & Brandt B (1959) [Allergic and toxic dermatitis in the manufacturing and processing of hexachlorocyclohexane.] Arch Gewerbepathol Gewerbehyg **17**: 365–383 (in German).

Benes V & Sram R (1969) Mutagenic activity of some pesticides in *Drosophila melanogaster*. Ind Med Surg **38**(12): 442–444.

Benezet HJ & Matsumura F (1973) Isomerization of γ-BHC to α-BHC in the environment. Nature **243**: 480–481.

Bercovici B, Wassermann M, Cucos S, Ron M, Wassermann D, & Pines A (1983) Serum levels of polychlorinated biphenyls and some organochlorine insecticides in women with recent and former missed abortion. Environ Res **30**: 169–174.

Bernhardt H & Ziemons E (1974) [Pesticide content of 19 German reservoirs. German Association of Gas and Water Experts.] Wasser **5**: 1–104 (in German).

Berry DH, Brewster MA, Watson R, & Neuberg RW (1987) Untoward effects associated with lindane abuse. Am J Dis Child **141**(2): 125–126.

Bertram H., Kemper FH, & Zenzen C (1980) [Occurrence of HCH isomers in man.] In: Deutsche Forschungsgemeinschaft, ed. [Hexachlorcyclohexane as a harmful substance in foods. Papers from two symposia of the Senate Committee on the Testing of Residues in Foods, held on 28–29 November 1979 and 6 March 1980.] Weinheim, Verlag Chemie, pp. 155–163 (in German).

Bidleman TF & Leonard R (1982) Aerial transport of pesticides over the northern Indian Ocean and adjacent seas. Atmos Environ **16**: 1099–1107.

Blok SMG, Greve PA, Sangster B, Savelkoul TJF, & Wegman RCC (1984) [Investigation of normally occurring values of a number of organochlorine pesticides and related compounds and their metabolites, of polychlorobiphenyls and of chlorophenols in the blood or plasma of healthy volunteers.] Bilthoven, National Institute of Public Health and Environmental Hygiene (Report No. 638101001) (in Dutch).

Blus LJ, Henny CJ, Lenhart DJ, & Kaiser TE (1984) Effects of heptachlor- and lindane-treated seed on Canada geese. J Wildl Manage **48**(4): 1097–1110.

References

Bluzat R & Seuge J (1979) Etude de la toxicité chronique de deux insecticides (carbaryl et lindane) de la generation F_1 de *Lymnea stagnalis* L. (mollusque gasteropode pulmone). I. Croissance des coquilles. Hydrobiologia **65**(3): 245–255.

Bosch AL (1987a) Dermal absorption of ^{14}C-lindane in male rabbits. Madison, Wisconsin, Hazleton Laboratories America, Inc. (Unpublished report No. 6188-103, submitted to WHO by CIEL).

Bosch AL (1987b) Dermal absorption of ^{14}C-lindane in male rats. Madison, Wisconsin, Hazleton Laboratories America, Inc. (Unpublished report No. 6188-104, submitted to WHO by CIEL).

Boulekbache H (1980) Study of the toxicity of the alpha-, beta-, and gamma isomers of hexachlorocyclohexane to fish, using the guppy as the test species. Paris, Université de Paris VII, Laboratoire d'Anatomie comparée (Unpublished report).

Boyd CE & Ferguson DE (1964) Susceptibility and resistance of mosquito fish to several insecticides. J Econ Entomol **57**: 430–431.

Boyd IL, Myhill DG, & Mitchell-Jones AJ (1988) Uptake of gamma-HCH (lindane) by Pipistrelle bats and its effect on survival. Environ Pollut **51**: 95–111.

Bradbury FR (1963) The systemic action of benzene hexachloride seed dressings. Ann Appl Biol **52**: 361–370.

Bradbury FR & Whitaker WO (1956) The systemic action of benzenehexachloride in plants: Quantitative measurements. J Sci Food Agric **7**: 248–253.

Brassow HL, Baumann K, & Lehnert G (1981) Occupational exposure to hexachlorocyclohexane. II. Health conditions of chronically exposed workers. Int Arch Occup Environ Health **48**: 81–87.

Bringmann G & Kühn R (1978) [Threshold values for the harmful effect of water-endangering substances on blue algae (*Microcystis aeruginosa*) and green algae (*Scenedesmus quadricauda*) in the cell reproduction inhibition test. Jahresber Wasser **50**: 45–60 (in German).

Brown D (1988) Lindane: 13-week dermal toxicity study (with interim kill and recovery period) in the rat (HUK Project No. 580/2). Harrogate, North Yorkshire, Hazleton, UH (Unpublished report submitted to WHO by CIEL).

Buscher CA, Dougherty JH, & Skrinde RJ (1964) Chemical oxidation of selected organic pesticides. J Water Pollut Control Fed **36**: 1005.

Buselmaier W, Röhrborn G, & Propping P (1972) Mutagenicity investigations with pesticides in the host-mediated assay and the dominant lethal test in mice. Biol Zbl **91**: 311–325 (in German).

Büsser MT & Lutz WK (1987) Stimulation of DNA synthesis in rat and mouse liver by various tumor promoters. Carcinogenesis **8**(10): 1433–1437.

Butler PA (1963a) Commercial fisheries investigations. Washington, DC, US Fish and Wildlife Service (Circular 199), pp. 5–28.

Butler PA (1963b) Commercial fisheries investigations. Washington, DC, US Fish and Wildlife Service (Circular 167), pp. 11–25.

Butler PA (1971) Influence of pesticides on marine ecosystems. Proc R Soc Lond **B177**: 321-329.

Cameron GR (1945) Risks to man and animals from the use of 2,2-bis(p-chlorophenyl),1,1,1-trichlorethane (DDT): with a note on the toxicology of γ-benzenehexachloride (666, gammexane). Br Med Bull **3**: 783–785.

Camon L, Martinez E, Artigas F, Sola C, & Rodriguez-Farre E (1988a) The effect of non-convulsant doses of lindane on temperature and body weight. Toxicology **49**: 389–394.

Camon L, Sola C, Martinez E, Sanfeliu C, & Rodriguez-Farre E (1988b) Cerebral glucose uptake in lindane-treated rats. Toxicology **49**: 381–387.

Carey AE, Gowen JA, Tai H, Mitchell G, & Wiersma GB (1979) Pesticide residue levels in soils and crops from 37 states, 1972–National Soils Monitoring Program (IV). Pestic Monit J **12**(4): 209–229.

Carrasco JM, Cunat P, Martinez M, & Primo E (1976) Pesticide residues in total diet samples, Spain 1971–1972. Pestic Monit J **10**(1): 18–23.

Cerey K, Szokolayova J, Rosival L, & Gencik A (1975) Detection of mutagenic actions of lindane by means of the dominant lethal test. In: VIII International Plant Protection Congress. Reports and information. Section IV: Plant protection in relation to human health and environmental pollution. Moscow, pp. 34–43.

Cetinkaya M, Gabel B, Podbielski A, & Thiemann W (1984) [On the correlation of nutrition and living habits of feeding mothers and contamination of human milk with nonvolatile chlorinated organic chemicals.] Akt Ernähr **9**(4): 157–162 (in German).

Chabert D & Vicente N (1978) Contamination de mollusques méditerranéens par un biocide organochloré: le lindane. Rev Int Océanog Méd **49**: 45–48.

Chadwick RW & Copeland MF (1985) Investigation of HCB as a metabolite from female rats treated daily for six days with lindane. J Anal Toxicol **9**: 262–266.

Chadwick RW & Freal JJ (1972a) The identification of five unreported lindane metabolites recovered from rat urine. Bull Environ Contam Toxicol **7**(2/3): 137–146.

Chadwick RW & Freal JJ (1972b) Comparative acceleration of lindane metabolism to chlorophenols by pretreatment of rats with lindane or with DDT and lindane. Food Cosmet Toxicol **10**: 789–795.

Chadwick RW, Cranmer MF, & Peoples AJ (1971) Comparative stimulation of γ-HCH metabolism by pretreatment of rats with γ-HCH, DDT, and DDT + γ-HCH. Toxicol Appl Pharmacol **18**: 685–695.

Chadwick RW, Chuang LT, & Williams K (1975) Dehydrogenation. A previously unreported pathway of lindane metabolism in mammals. Pestic Biochem Physiol **5**: 575–586.

Chadwick RW, Freal JJ, Sovocool GW, Bryden CC, & Copeland MF (1978) The identification of three previously unreported lindane metabolites recovered from mammals. Chemosphere **8**: 633–640.

Chadwick RW, Copeland MF, Mole ML, Nesnow S, & Cooke N (1981) Comparative effect of pretreatment with phenobarbital, Arochlor 1254, and β-naphthaflavone on the metabolism of lindane. Pestic Biochem Physiol **15**: 120–136.

Chadwick RW, Copeland MF, Wolff GL, Stead AG, Mole ML, & Whitehouse DA (1987) Saturation of lindane metabolism in chronically treated (YS x VY)F$_1$ hybrid mice. J Toxicol Environ Health **20**: 411–434.

References

Chadwick RW, Cooper RL, Chang J, Rehnberg GL, & McElroy WK (1988) Possible antiestrogenic activity of lindane in female rats. J Biochem Toxicol **3**: 147–158.

Chakravarty S, Mandal A, & Lahiri P (1986) Effect of lindane on clutch size and level of egg protein in domestic duck. Toxicology **39**: 93–103.

Charnetski WA & Lichtenstein EP (1973) Penetration and translocation of 14C-lindane in pea plants. J Econ Entomol **66**(2): 344–349.

Chattopadhyay P, Karnik AB, Thakore KN, Lakkad BC, Nigam SK & Kashyap SK (1988) Health effects among workers involved in the manufacture of hexachlorocyclohexane. J Soc Occup Med **38**: 77–81.

Chen C-P (1968) The effect of a protein-deficient diet on the acute oral toxicity of lindane. Kingston, Ontario, Queen's University (Thesis).

Chen T & Liang C (1956) Oral toxicity of lindane and its tolerance in poultry and mice. J Agric Assoc China **15**: 78–90.

Chessells MJ, Hawker DW, Connell DW, & Papajcsik IA (1988) Factors influencing the distribution of lindane and isomers in soil of an agricultural environment. Chemosphere **17**(9): 1741–1749.

Clausing P, Grün G, & Beitz H (1980) [Possibilities for investigating and avoiding damage to bird life by pesticides.] Nachrichtenbl Pflanzenschutzdienst DDR **34**(7): (in German).

Cliath MM & Spencer WF (1971) Division S-6-soil and water management and conservation. Soil Sci Soc Am Proc **35**: 791–795.

Cliath MM & Spencer WF (1972) Dissipation of pesticides from soil by volatilization of degradation products. I. Lindane and DDT. Environ Sci Technol **6**(10): 910–914.

Codex Alimentarius Commission (1986) Guide to Codex recommendations concerning pesticide residues, Part 2. Maximum limits for pesticide residues, 3rd prelim. issue, Rome, Food and Agricultural Organization of the United Nations (CAC/Vol. 8-Ed 2-1986).

Cope OB (1965) Effects of pesticides on fish and wildlife (1964 research finding). Washington, DC, US Fish and Wildlife Service (Circular 226), pp. 51–63.

Copeland MF & Chadwick RW (1979) Bioisomerization of lindane in rats. J Environ Pathol Toxicol **2**: 737–749.

Coper H, Herken H, & Klempau I (1951) [Antagonism and synergism of β- and γ-hexachlorocyclohexane.] Klin Wochenschr **29**(13/14): 264–265 (in German).

Corneliussen PE (1969) Residues in food and feed. Pesticide residues in total diet samples (IV). Pestic Monit J **2**(4): 140–152.

Cosson Mannevy MA & Marchand MH (1980) Bioaccumulation des éléments chimiques chez les organismes marins: limites des études *in vitro*. Exposé fait au symposium de Valbonne, 1–2 October 1980. (Unpublished report Celamerck C80 P007C submitted to WHO by CIEL).

Cowan AA (1981) Organochlorine compounds in mussels from Scottish coastal waters. Environ Pollut **B2**: 129–143.

Culley DD Jr & Ferguson DE (1969) Patterns of insecticide resistance in the mosquito fish, (*Gambusia affinis*). J Fish Res Board Can **26**: 2395–2401.

Cummings JG, Zee KT, Turner V, & Quinn F (1966) Residues in eggs from low level feeding of five chlorinated hydrocarbon insecticides to hens. J Assoc Off Anal Chem **49**(2): 354-364.

Czeglédi-Janko G & Avar P (1970) Occupational exposure to lindane; clinical and laboratory findings. Br J Ind Med **27**: 283–286.

Dahlen JH & Haugen AO (1954) Acute toxicity of certain insecticides to the bobwhite quail and mourning dove. J Wildl Manage **18**: 477–481.

Darskus R (1982) Water solubility and partition coefficient *n*-octanol/water for CM active ingredients. (Unpublished report Celamerck 111AC-114-006 submitted to WHO by CIEL).

Das B & Singh PK (1977) Detoxication of the pesticide benzenehexachloride by blue-green algae. Microbios Lett **4**: 99–102.

Davies D & Mes J (1987) Comparison of the residue levels of some organochlorine compounds in breast milk of the general and indigenous Canadian populations. Bull Environ Contam Toxicol **39**: 743–749.

Davies JE, Dedhia HV, Morgade C, Barquet A, & Maibach HI (1983) Lindane poisoning. Arch Dermatol **119**(2): 142–144.

Davis HC & Hidu H (1969) Effects of pesticides on embryonic development of clams and oysters and on survival and growth of larvae. Fish Bull **67**: 393–404.

De Brabander M, van de Veire R, Aerts F, Geuens S, & Hoebeke J (1976) New culture model facilitating rapid quantitative testing of mitotic spindle inhibition in mammalian cells. J Natl Cancer Inst **56**(2): 357–363.

De Bruijn J (1979) Reduction of chlordane, DDT, heptachlor, hexachlorobenzene and hexachlorocyclohexane isomers contained in effluents. Brussels, Commission of the European Communities (Report ENV/223/74-E Rev.2).

Deichmann WB & MacDonald WE (1977) Organochlorine pesticides and liver cancer deaths in the United States, 1930–1972. Ecotoxicol Environ Saf **1**: 89–110.

Demozay D & Marechal G (1972) Physical and chemical properties. In: Ulmann, E., ed., Lindane: Monograph of an insecticide. Freiburg im Breisgau, K. Schillinger Verlag, pp. 14–21.

Deo PG, Hasan SB, & Majumder SK (1981) Interconversions and toxicity changes in hexachlorocyclohexane isomers on dispersion in water. J Environ Sci Health **B16**(6): 691–701.

Desi I (1972) In: Proceedings of the International Symposium on Lindane, Vienna, 9 June. Budapest, National Institute of Public Health. (Unpublished report submitted to WHO by CIEL).

Desi I (1974) Neurotoxicological effect of small quantities of lindane. Int Arch Arbeitsmed **33**: 152–162.

Desi I (1976) Lindane: Toxicological studies. In: Proceedings of the Symposium on Lindane, Lyon-Chazay, 29 April 1976. Brussels, Centre International d'Etudes du Lindane, pp. 67–69.

Desi I (1983) Neurotoxicological investigation of pesticides in animal experiments. Neurobehav Toxicol Teratol **5**: 503–515.

References

Desi I, Dura G, & Szolobodnyik J (1977) Testing of pesticide toxicity in tissue culture. J Toxicol Environ Health 2: 1053–1066.

Desi I, Varga L, & Farkas I (1978) Studies on the immunosuppressive effect of organochlorine and organophosphoric pesticides in subacute experiments. J Hyg Epidemiol Microbiol Immunol 22(1): 115–122.

Deutsche Forschungsgemeinschaft (1978) [Residues in breast milk—situation and assessment]. Boppard, Harald Boldt Verlag (in German).

Deutsche Forschungsgemeinschaft (1979) [Residue analysis of pesticides (5th instalment): Organochlorine pesticides]. Weinheim, Verlag Chemie, pp. 12–18 (in German).

Deutsche Forschungsgemeinschaft (1983) [Hexachlorcyclohexane as a harmful substance in foods. Papers from two symposia of the Senate Committee on the Testing of Residues in Foods, held on 28–29 November 1979 and 6 March 1980]. Weinheim, Verlag Chemie (in German).

Dewitt JB, Stickel WH, & Springer PF (1963) Wildlife studies, Patuxent Wildlife Research Center, 1961–1962. US Fish and Wildlife Service (Circular 167), pp. 74–96.

van Dijck P & van de Voorde H (1976) Mutagenicity versus carcinogenicity of organochlorine insecticides. Meded Fac Landbouwwet Rijksuniv Gent 41(2): 1491–1498.

Dion BM (1984) Toxicité aigue de l'isomère γ de l'hexachlorocyclohexane à l'égard des truites arc-en-ciel et fario. (Unpublished report Celamerck 111AC-442-006 submitted to WHO by CIEL).

Dittrich V (1966) Investigation on the acute oral toxicity of different pesticides on nestlings of *Parus ater* and adults of *Passer domesticus*. Z Angew Entomol 57: 430–437.

Doisy EA & Bocklage BC (1949) Chronic toxicity of gamma isomer of hexachlorocyclohexane in the albino rat. Proc Soc Exp Biol Med 71: 490–493.

Doisy EA & Bocklage BC (1950) Inositol and the toxicity of four isomers of benzenehexachloride for the rat. Proc Soc Exp Biol Med 74: 613–616.

Drummond L, Gillanders EM, & Wilson HK (1988) Plasma γ-hexachlorocyclohexane concentrations in forestry workers exposed to lindane. Br J Ind Med 45: 493–497.

Duee PH, Bories G, Froc J, Hascoet M, Henry Y, Peleran JC, & Conseil G (1975) Influence d'une teneur élevée en pesticide (lindane) dans l'aliment sur le taux d'ovulation et la mortalité embryonnaire chez la truie. J Rech Porcine Fr: 331–335.

Dugast P (1980) Cheminement du lindane dans une châine trophique terrestre de laboratoire. Meded Fac Landbouwwet Rijksuniv Gent 45(4): 905–914.

Duggan RE & Corneliussen PE (1972) Dietary intake of pesticide chemicals in the United States (III), June 1968–April 1970. Pestic Monit J 5(4): 331–341.

Dutch Chemical Industry Association (1980) 1α, 2α, 3β, 4α, 5α, 6β-hexachlorocyclohexane. In: Handling chemicals safely, 2nd ed. Amsterdam, Dutch Association of Safety Experts and Dutch Safety Institute, p. 529.

Dzwonkowska A & Hubner H (1986) Induction of chromosomal aberrations in the Syrian hamster by insecticides tested *in vivo*. Arch Toxicol 58: 152–156.

Earl FL, Miller E & van Loon EJ (1973) Reproductive, teratogenic and neonatal effects of some pesticides and related compounds in beagle dogs and miniature swine. In: Deichmann WB, ed., Pesticides and environment: a continuing controversy. New York, Intercontinental Medical Book Corporation, pp. 253–266.

Eckenhausen FW, Bennett D, Beynon KI, & Elgar KE (1981) Organochlorine pesticide concentrations in perinatal samples from mothers and babies. Arch Environ Health 36(2): 81–92.

Edelman T (1984) [Background values of a number of inorganic and organic substances in the soil of the Netherlands; an initial reconnaissance]. The Hague, State Publishing House (Soil protection, report VROM 34) (in Dutch).

Eder G, Sturm R, & Ernst W (1987) Chlorinated hydrocarbons in sediments of the Elbe River and the estuary. Chemosphere 16(10–12): 2487–2496.

Edson EF, Sanderson DM, & Noakes DN (1966) Acute toxicty data for pesticides. World Rev Pestic Control 5(3): 143–151.

Edwards CA (1966) Insecticide residues in soils. Residue Rev 13: 83–132.

Edwards CA (1973a) Persistent pesticides in the environment. Cleveland, Ohio, CRC Press.

Edwards CA, ed. (1973b) Environmental pollution by pesticides. New York, Plenum Press.

Edwards CA (1977) Environmental aspects of the usage of pesticides in developing countries. Med. Fac. Landbouwwet Rijksuniv Gent 42/2: 853–868.

Edwards CA (1981) Lindane exposure analysis (August 1980). in: Response of the Centre International d'Etudes du Lindane to EPA's preliminary Notice of Determination and Position Document 2/3 on lindane. Brussels, Centre International d'Etudes du Lindane, Vol. 1.

Egan H & Hubbard AW (1975) Analytical surveys of food. Br Med Bull 31(3): 201–208.

Ehlers W, Farmer WJ, Spencer WF, & Letey J (1969a) Lindane diffusion in soils. II. Water content, bulk density and temperature effects. Soil Sci Soc Am Proc 33: 505–508.

Ehlers W, Letey, J, Spencer WF, & Farmer WJ (1969b) Lindane diffusion in soils. I. Theoretical consideration and mechanism of movement. Soil Sci Soc Am Proc 33: 501–504.

Eichler D (1975) Investigations of residue samples for possible metabolites of lindane. (Unpublished report Celamerck 111AA-641-003 submitted to WHO by CIEL).

Eichler D (1977) Experiments on the decomposition of lindane when exposed to UV-light. (Unpublished report Celamerck 111AC-143-002 submitted to WHO by CIEL).

Eichler D (1980) [Biotic and abiotic breakdown and conversion behaviour, including isomerization (plant).]. In: Deutsche Forschungsgemeinschaft, ed. [Hexachlorcyclohexane as a harmful substance in foods. Papers from two symposia of the Senate Committee on the Testing of Residues in Foods, held on 28–29 November 1979 and 6 March 1980]. Weinheim, Verlag Chemie, pp. 65–72 (in German).

Eichler D, Heupt W, & Fischer O (1976) [Attempts at lindane breakdown in biomes]. (Unpublished report Celamerck no. 111AX-912002 submitted to WHO by CIEL) (in German).

References

Eichler D, Heupt W, & Paul W (1983) Comparative study on the distribution of α- and γ-hexachlorocyclohexane in the rat with particular reference to the problem of isomerization. Xenobiotica 13(11): 639–647.

Eisler R (1969) Acute toxicities of insecticides to marine decapod crustaceans. Crustaceana Int J Crustacean Res 16(3): 302–310.

Eisler R (1970a) Latent effects of insecticide intoxication to marine molluscs. Hydrobiologia 36(3–4): 345–352.

Eisler R (1970b) Acute toxicities of organochlorine and organophosphorus insecticides to estuarine fishes. Washington, DC, US Department of Interior, Bureau of Sport Fisheries and Wildlife (Tech. paper 45), pp. 3-12.

Eisler R (1970c) Factors affecting pesticide-induced toxicity in an estuarine fish. Washington DC, US Department of Interior, Bureau of Sport Fisheries and Wildlife (Technical paper 45), pp. :2-20.

Eldefrawi ME, Sherby SM, Abalis IM, & Eldefrawi AT (1985) Interaction of pyrethroid and cyclodiene insecticides with nicotinic acetylcholine and GABA receptors. Neurotoxicology 6: 47–62.

El-Dib MA & Badawy MI (1985) Organochlorine insecticides and PCBs in water, sediment and fish from the Mediterranean Sea. Bull Environ Contam Toxicol 34: 216–227.

Elsner E, Bieniek D, Klein W, & Korte F (1972) [Contributions to ecological chemistry. LII. Distribution and conversion of aldrin-^{14}C, heptachlor-^{14}C and lindane-^{14}C in the green alga *Chlorella pyrenoidosa*.] Chemosphere 6: 247–250 (in German).

Engst R, Blazovich M, & Knoll R (1967) [Occurrence of lindane in carrots and its influence on the carotene content]. Nahrung 11(5): 389–399 (in German).

Engst R, Macholz RM, & Kujawa M (1970) Recent state of lindane metabolism. Residue Rev 68: 59–84.

Engst R, Macholz RM, & Kujawa M (1974) [Lindane metabolism. Degradation of lindane by mould cultures. Unconjugated metabolites]. Nahrung 18(8): 737–745 (in German).

Engst R, Macholz RM, Kujawa M, Lewerenz HJ, & Plass R (1976) The metabolism of lindane and its metabolites gamma-2,3,4,5,6-pentachlorocyclohexene, pentachlorobenzene and pentachlorophenol in rats and the pathways of lindane-metabolism. J Environ Sci Health B11(2): 95–117.

Engst R, Macholz RM, & Kujawa M (1977) The metabolism of lindane in a culture of mould and the degradation scheme of lindane. Chemospere 7: 401–418.

Engst R, Macholz RM, & Kujawa M (1978a) [Confirmations of the degradation scheme of gamma-hexachlorocyclohexane]. Nahrung 22(6): K29–K32 (in German).

Engst R, Macholz RM, & Kujawa M (1978b) [Metabolites of hexachlorocyclohexane (HCH) isomers in human blood]. Pharmazie 33(2/3): 109–111 (in German).

Engst R, Macholz RM, & Kujawa M (1979a) Recent state of lindane metabolism, Part II. Residue Rev 75: 71–95.

Engst R, Fritsche W, Knoll R, Kujawa M, Macholz RM, & Straube G (1979b) Interim results of studies of microbial isomerization of gamma-hexachlorocyclohexane. Bull Environ Contam Toxicol 22: 699–707.

Epstein SS, Arnold E, Andrea J, Bass W, & Bishop Y (1972) Detection of chemical mutagens by the dominant lethal assay in the mouse. Toxicol Appl Pharmacol 23: 288–325.

Ernst W (1975) [Uptake, excretion and conversion of lindane-^{14}C by *Mytilus edulis* (common mussel)]. Chemosphere 6: 375–380 (in German).

Ernst W (1979) Factors affecting the evaluation of chemicals in laboratory experiments using marine organisms. Ecotoxicol Environ Saf 3: 90–98.

Faladysz J & Szefer P (1982) Chlorinated hydrocarbons in diving ducks wintering in Gdansk Bay, Baltic Sea. Sci Total Environ 24: 119–127.

FAO (1973) Organochlorine insecticides 1973 (BHC, BHC + DDT, γ-BHC, campheclor, chlordane, DDT, endosulfan, endrin, heod, heptachlor, HHDN). Rome, Food and Agricultural Organization of the United Nations, pp. 29–37 (FAO Specifications for Pesticides).

FAO/WHO (1967) Evaluation of some pesticide residues in food. Geneva, World Health Organization (FAO: PL/CP/15; WHO Food Add./67.32).

FAO/WHO (1968) 1967 Evaluation of some pesticide residues in food. Geneva, World Health Organization (FAO: PL:1967/M/11/1; WHO Food Add./68.30).

FAO/WHO (1969) 1968 Evaluation of some pesticide residues in food. Geneva, World Health Organization (FAO: PL:1968/M/9/1; WHO Food Add./69.35).

FAO/WHO (1970) 1969 Evaluations of some pesticide residues in food. Geneva, World Health Organization (FAO: PL:1969/M/17/1; WHO Food Add./70.38).

FAO/WHO (1974) 1973 Evaluations of some pesticide residues in food. Geneva, World Health Organization (AGP: 1973/M/9/1; WHO Pesticide Residues Series No. 3).

FAO/WHO (1975) 1974 Evaluations of some pesticide residues in food. Geneva, World Health Organization (AGP: 1974/M/11; WHO Pesticide Residues Series No. 4).

FAO/WHO (1976) 1975 Evaluations of some pesticide residues in food. Geneva, World Health Organization (AGP: 1975/M/13; WHO Pesticide Residues Series No. 5).

FAO/WHO (1978) 1977 Evaluations of some pesticide residues in food. Rome, Food and Agriculture Organization of the United Nations (FAO Plant Production and Protection Paper 10 Sup).

FAO/WHO (1980) 1979 Evaluations of some pesticide residues in food. Rome, Food and Agriculture Organization of the United Nations (FAO Plant Production and Protection Paper 20 Sup).

FAO/WHO (1990) Pesticide residues in food. Report of the 1989 Joint Meeting of the FAO Panel of Experts on Pesticide Residues in Food and the Environment and the WHO Expert Group on Pesticide Residues. Rome, Food and Agriculture Organization of the United Nations (FAO Plant Production and Protection Paper 99).

Feldmann RJ & Maibach HI (1974) Percutaneous penetration of some pesticides and herbicides in man. Toxicol Appl Pharmacol 28: 126–132.

Fieggen W (1983) [Report on investigation of purification mud on pesticides and PCBs in 1981]. The Hague, Union of Polder Boards (in Dutch).

References

Fishman BE & Gianutsos G (1987) Opposite effects of different hexachlorocyclohexane (lindane) isomers on cerebellar cyclic GMP: relation of cyclic GMP accumulation to seizure activity. Life Sci 41: 1703–1709.

Fishman BE & Gianutsos G (1988) CNS biochemical and pharmacological effects of the isomers of hexachlorocyclohexane (lindane) in the mouse. Toxicol Appl Pharmacol 93: 146–153.

Fitzhugh OG, Nelson AA, & Frawley JP (1950) The chronic toxicities of technical benzenehexachloride and its alpha, beta- and gamma isomers. J Pharm Exp Ther 100(1): 59–66.

Fitzloff JF & Pan JC (1984) Epoxidation of the lindane metabolite, γ-PCCH, by human and rat liver microsomes. Xenobiotica 14(7): 599–604.

Fitzloff J., Portig J, & Stein K (1982) Lindane metabolism by human and rat liver microsomes. Xenobiotica 12(3): 197–202.

Fooken C & Butte W (1987) Organochlorine pesticides and polychlorinated biphenyls in human milk during lactation. Chemosphere 16(6): 1301–1309.

Foulquier L, Reynier B, Granby A, & Bovard P (1971) Toxicité du lindane sur l'anguille, France. CR Acad Agric 57(12): 1060–1068.

Francis AJ, Spanggord RJ, & Ouchi GI (1975) Degradation of lindane by *Escherichia coli*. Appl Microbiol 29(4): 567–568.

Frank R, Braun HE, Sirons GH, Rasper J, & Ward GG (1985) Organochlorine and organophosphorus insecticides and industrial pollutants in the milk supplies of Ontario, 1983. J Food Prod 48(6): 499–504.

Franklin A (1987) The concentration of metals, organochlorine pesticide and PCB residues in marine fish and shell fish: results from MAFF fish and shell fish monitoring programmes, 1977-1984. Lowestoft, Ministry of Agriculture, Fisheries and Food, Directorate of Fisheries Research (Aquatic Environment Monitoring Report No. 16).

Freal JJ & Chadwick RW (1973) Metabolism of hexachlorocyclohexane to chlorophenols and effect of isomer pretreatment on lindane metabolism in rat. J Agric Food Chem 21(3): 424–427.

Fricke G (1972) [Contamination of soils with chlorinated hydrocarbon insecticides, comparison of 1969 with 1972 (Communication 1, field vegetable growing). Gesunde Pflanz 24(11): 177–179 (in German).

Frisque GE, Galoux M, & Bernes A (1983) Accumulation de deux micropollutants (les polychlorobiphenyles et le gamma-HCH) par des bryophytes aquatiques de la Meuse. Meded Fac Landbouwwet Rijksuniv Gent 48(4): 971–983.

Frohberg H & Bauer A (1972a) Lindane. Testing for teratogenic effects in mice following subcutaneous injection. Darmstadt, E. Merck (Unpublished report 4/24/72 submitted to WHO by CIEL).

Frohberg H & Bauer A (1972b) Lindane. Testing for teratogenic effects in mice following oral administration. Darmstadt, E. Merck (Unpublished report 4/107/72 submitted to WHO by CIEL).

Frohberg H & Bauer A (1972c) [Lindane: Testing for mutagenic effect. Dominant lethality test on male mice]. Darmstadt, E. Merck (Unpublished report 4/8/72 submitted to WHO by CIEL).

Frohberg H, Wolf HP, & von Eberstein M (1972a) [Testing for acute toxicity in rats after oral administration, intraperitoneal injection and intramuscular injection]. Darmstadt, E. Merck (Unpublished report submitted to WHO by CIEL) (in German).

Frohberg H, von Eberstein M, Engemann J, & Weisse G (1972b) [Testing for acute toxicity in rats after oral administration and intraperitoneal injection]. Darmstadt, E. Merck (Unpublished report submitted to WHO by CIEL) (in German).

Fuhremann TW & Lichtenstein EP (1980) A comparative study of the persistence, movement, and metabolism of six carbon-14 insecticides in soils and plants. J Agric Food Chem 28: 446–452.

Gaggi C, Bacci E, Calamari D, & Fanelli R (1986) Chlorinated hydrocarbons in plant foliage: An indication of the tropospheric contamination level. Chemosphere 14(11–12): 1673–1686.

Gaines TB (1960) The acute toxicity of pesticides to rats. Toxicol Appl Pharmacol 2: 88–99.

Gardais M & Scherrer M (1979) Photodecomposition of lindane (studies realized in Rhône-Poulenc (CRD)). Lyon, Rhône-Poulenc (Unpublished report submitted to WHO by CIEL).

Gartrell MJ, Craun JC, Podrebarac DS, & Gunderson EL (1985a) Pesticides, selected elements and other chemicals in adult total diet samples, October 1978–September 1979. J Assoc Off Anal Chem 68(5): 862–875.

Gartrell MJ, Craun JC, Podrebarac DS, & Gunderson EL (1985b) Pesticides, selected elements and other chemicals in infant and toddler total diet samples, October 1978–September 1979. J.Assoc Off Anal Chem 68(5): 842–861.

Garretson AL & San Clemente CL (1968) Inhibition of nitrifying chemolithotrophic bacteria by several insecticides. J Econ Entomol 61(1): 285–288.

Gaur AC & Misra KC (1977) Effect of simazine, lindane and ceresan on soil respiration and nitrification rates. Plant Soil 46: 5–15.

Gencik A (1977) [Cytogenic examination of the bone marrow of rats after administration of lindane]. Bratislava Lek Listy 67(5): 579–582 (in Slovak).

Geyer H, Sheehan P, Kotzias D, Freitag D, & Korte F (1982) Prediction of ecotoxicological behaviour of chemicals. Relationship between physico-chemical properties and bioaccumulation of organic chemicals in the mussel *Mytilus edulis*. Chemosphere 11(11): 1121–1134.

Geyer H, Politzki G, & Freitag D (1984) Prediction of ecotoxicological behaviour of chemicals. Relationship between n-octanol/water partition coefficient and bioaccumulation of organic chemicals by alga *Chlorella*. Chemosphere 13(2): 269–284.

Geyer H, Scheunert I, & Korte F (1986) Bioconcentration potential of organic environmental chemicals in humans. Regul Toxicol Pharmacol 6: 313–347.

Ginsburg CM & Lowry W (1983) Absorption of gamma benzene hexachloride following application of Kwell shampoo. Pediatr Dermatol 1(1): 74–76.

Ginsburg CM, Lowry W, & Reisch J (1977) Absorption of lindane (gamma benzene hexachloride) in infants and children. J Pediatr 91(6): 998-1000.

References

Glatt HR & Oesch F (1984) Mammalian cell (V79) mutagenicity test on lindane. Brussels, CIEL (Unpublished report Celamerck No. 111AC-457-019, submitted to WHO by CIEL).

Gopalswamy UV & Aiyar AS (1984) Biotransformation of lindane in the rat. Bull Environ Contam Toxicol **32**: 148–156.

Gorchev HG & Jelinek CF (1985) A review of the dietary intakes of chemical contaminants. World Health Organ Bull **63**(5): 945–962.

Goto M, Hattori M, & Miyagawa T (1972) [Contribution to ecological chemistry. II. Hepatoma formation in mice after administration of high doses of HCH isomers]. Chemosphere **6**: 279-282 (in German).

Graeve K & Herrnring G (1949) [The application of the gamma isomers of hexachlorocyclohexane as an anthelmintic]. Klin Wochenschr **27**: 318 (in German).

Graeve K & Herrnring G (1951) [The toxicity of gamma-hexachlorocyclohexane]. Arch int pharmacodyn **85**(1/2): 64-72 (in German).

Green DWJ, Kendall A, Williams A, & Pascoe D (1986) Studies on the acute toxicity of pollutants to freshwater macroinvertebrates. 4. Lindane (γ-hexachlorocyclohexane). Arch Hydrobiol **106**(2): 263–273.

Greve PA (1972) Potentially hazardous substances in surface waters. Part I. Pesticides in the River Rhine. Sci Total Environ **1**: 173–180.

Greve PA & van Harten DC (1983) [Relationship between levels of organochlorine pesticides and PCBs in fat and blood in human subjects]. Bilthoven, National Institute of Public Health and Environmental Hygiene (Unpublished report No. 638219001) (in Dutch).

Greve PA & van Hulst S (1977) [Organochlorine pesticides and PCBs in total diets]. Bilthoven, National Institute of Public Health and Environmental Hygiene (Unpublished report No. 192/77 Tox-ROB) (in Dutch).

Greve PA & Wegman RCC (1985) Organochlorine compounds in human milk; data from a recent investigation in the Netherlands. Working paper for the WHO consultation on organochlorine compounds in human milk and related hazards, Bilthoven, 9–11 January 1985. Bilthoven, Institute of Public Health and Environmental Hygiene.

Grover PL & Sims P (1965) The metabolism of γ-2, 3,4,5,6-pentachlorocyclo-1-ene and γ-hexachlorocyclohexane in rats. Biochem J **96**: 521–525.

Guenard J, Eichelberger HP, & Terrier C (1984a) In vivo sister chromatid exchange assay in CF_1-mouse bone marrow cells with lindane (oral application). Itingen, Switzerland, Research and Consulting Company AG. (Unpublished report No. 025705, Celamerck No. 111AC-457-020, submitted to WHO by CIEL).

Guenard J, Eichelberger HP, & Terrier C (1984b) *In vivo* sister chromatid exchange assay in CF_1 mouse bone marrow cells with lindane (intraperitoneal injection). Itingen, Switzerland, Research and Consulting Company AG. (Unpublished report No. 025716, Celamerck No. 111AC-457-021, submitted to WHO by CIEL).

Guenzi WD & Beard WE (1970) Volatilization of lindane and DDT from soils. Soil Sci Soc Am Proc **34**: 443–447.

Guicherit R & Schulting FL (1985) The occurrence of organic chemicals in the atmosphere of the Netherlands. Sci Total Environ **43**: 193–219.

Gunderson EL (1988) FDA Total Diet Study, April 1982–April 1984, dietary intakes of pesticides, selected elements and other chemicals. J Assoc Off Anal Chem **71**(6): 1200–1209.

Gupta SK, Parikh JR, Shah M, Chatterjee SK & Kashyap SK (1982) Changes in serum HCH residues in malaria spraymen after short-term occupational exposure. Arch Environ Health **37**(1): 41–44.

Haider K (1979) Degradation and metabolization of lindane and other hexachlorocyclohexane isomers by anaerobic and aerobic soil microorganisms. Z Naturforsch **34C**: 1066–1069.

Haider K (1983) [Breakdown and conversion of gamma-HCH and other HCH isomers by soil microorganisms]. In: [Hexachlorcyclohexane as a harmful substance in foods]. Bonn, German Research Council, pp. 73–78 (DFG research report) (in German).

Haider K & Jagnow G (1975) [Degradation of ^{14}C-, ^{3}H- and ^{36}Cl-labelled γ-hexachlorocyclohexan by anaerobic soil microorganisms]. Arch Microbiol **104**(2): 113–121 (in German).

Haider K, Jagnow G, Kohnen R, & Lim SU (1974) [Degradation of chlorinated benzenes, phenols and cyclohexane derivatives by benzene and phenol using soil bacteria under aerobic conditions]. Arch Microbiol **96**: 183–200 (in German).

Haider K, Jagnow, G, & Rohr K (1975) [Anaerobic degradation of γ-hexachlorocyclohexane by a mixed bacterial flora from arable soils and from the cow rumen]. Landwirtsch Forsch **32**(2): 147–152 (in German).

Haider K, Jagnow G, & Rohr R (1976) [Anaerobic breakdown of γ-hexachlorocyclohexane by mixed bacterial flora from soil and cow's rumen]. Landwirtsch Forsch **32**(11): 147–152 (in German).

Hamada M, Kawano E, Kawamura S, & Shiro M (1981) Radiation- and photo-induced degradation of five isomers of 1,2,3,4,5,6-hexachlorocyclohexane. Agric Biol Chem **45**(3): 659–665.

Hamada M, Kawano E, Kawamura S, & Shiro M (1982) A new isomer of 1,2,3,4,5-pentachlorocyclohexane from UV-irradiation products of α-, β-, and δ-isomers of 1,2,3,4,5,6-hexachlorocyclohexane. Agric Biol Chem **46**(1): 153–157.

Hanada M, Yutani C, & Miyaji T (1973) Induction of hepatoma in mice by benzenehexachloride. Gann **64**(5): 511–513.

Hanig JP, Yoder PD, & Krop S (1976) Convulsions in weanling rabbits after a single topical application of 1% lindane. Toxicol Appl Pharmacol **38**: 463–469.

Hansen PD (1980) Uptake and transfer of the chlorinated hydrocarbon lindane (γ-BHC) in a laboratory freshwater food chain. Environ Pollut (Ser A), **21**: 97–108.

Harrison DL, Poole WSH, & Mol JCM (1963) Observations on feeding lindane-fortified mash to chickens. N Z Vet J **11**(6): 137–140.

Hartley DM & Johnston JB (1983) Use of the freshwater clam *Corbicula manilensis* as a monitor for organochlorine pesticides. Bull Environ ContamToxicol **31**: 33–40.

Hashimoto Y & Nishiuchi Y (1981) [Establishment of bioassay methods for the evaluation of acute toxicity of pesticides to aquatic organisms]. J Pestic Sci (Jpn), **6**: 257–264 (in Japanese).

References

Hawkins GS & Reifenrath WG (1984) Development of an *in vitro* model for determining the fate of chemicals applied to skin. Fundam Appl Toxicol **4**: 133–144.

Hawkinson JE, Shull LR, & Joy RM (1989) Effects of lindane on calcium fluxes in synaptosomes. Neurotoxicology **10**: 29–40.

Haworth S, Lawlor T, Mortelmans K, Speck W, & Zeiger E (1983) *Salmonella* mutagenicity test results for 250 chemicals. Environ Mutagen **Suppl 1**: 3–142.

Hayes WJ Jr (1982) Benzene hexachloride. In: Hayes WJ Jr (ed.) Pesticides studied in man. Baltimore, Williams and Wilkins, pp. 211–228.

Hazleton Laboratories America Inc (1976a) Teratology studies in rats—lindane (gamma benzene hexachloride, USP). Madison, Wisconsin. (Unpublished report submitted to WHO by CIEL).

Hazleton Laboratories America Inc (1976b) Teratology studies in rabbits—lindane (gamma benzene hexachloride, USP). Madison, Wisconsin. (Unpublished report submitted to WHO by CIEL).

Heeschen W, Nyhuis H & Blüthgen, A (1980) [Investigations on the significance of the transformation and degradation of benzene hexachloride (BHC) in the lactating cow and in the surroundings for the BHC-contamination of milk]. Milchwissenschaft **35**(4): 221–224 (in German).

Henderson C, Pickering QH, & Tarzwell CM (1959) Relative toxicity of ten chlorinated hydrocarbon insecticides to four carbon insecticides to species of fish. Trans Am Fish Soc **88**: 23–32.

Herbst M (1976) Toxicology of lindane. In: Proceedings of the symposium on lindane, Lyon-Chazay, 29 April 1976. Brussels, CIEL, pp. 43–69.

Herbst M, Weisse I, & Koel LMER (1975) A contribution to the question of the possible hepatocarcinogenic effects of lindane. Toxicology **4**: 91–96.

Heritage AD & MacRae LC (1977a) Identification of intermediates formed during the degradation of hexachlorocyclohexanes by *Clostridium sphenoïdes*. Appl Environ Microbiol **33**(6): 1295–1297.

Heritage AD & MacRae LC (1977b) Degradation of lindane by cell-free preparations of *Clostridium sphenoïdes*. Appl Environ Microbiol **34**(2): 222–224.

Heritage AD & MacRae LC (1979) Degradation of hexachlorocyclohexanes and structurally related substances by *Clostridium sphenoïdes*. Aust J Biol Sci **32**: 493–500.

Herken H (1950a) [Functional disorders of the central nervous system caused by hexachlorocyclohexane]. Naunyn-Schmiedeberger Arch Exp Pathol Pharmacol **212**: 158–159 (in German).

Herken H (1950b) [Functional disorders of the central nervous system following the action of hexachlorocyclohexane]. Arztliche Wochenschr **5**(13/14): 193–195 (in German).

Herken H (1950c) [Alterations in cell function in the nervous system caused by hexachlorocyclohexane]. Naunyn-Schmiedeberger Arch Exp Pathol Pharmacol **211**: 143–152 (in German).

Herken H (1951) [Disruption of the nervous system with hexachlorcyclohexane]. Arzneimittelforschung **1**: 356–359 (in German).

Hermens J & Leeuwangh P (1982) Joint toxicity of mixtures of 8 and 24 chemicals to the guppy (*Poecilia reticulata*). Ecotoxicol Environ Saf **6**: 302–310.

Herrmann R, Thomas W, & Huebner D (1984) [Behaviour of organic micropollutants (PAH, PCB and BHC) and of a fecal sterol in the Husum estuary and in the adjacent North Frisian Wadden Sea]. Dtsch Gewässerkd Mitt **4**: 101–107 (in German).

Herve S, Heinonen P, Paukku R, Knuutila M, Koistinen J, & Paasivirta J (1988) Mussel incubation method for monitoring organochlorine pollutants in water courses. Four-year application in Finland. Chemosphere **17**(10): 1945–1961.

Heupt W (1974) [Seepage behaviour of pesticides]. (Unpublished report Celamerck-Ingelheim No. 11110-922-001–11110-922-009, submitted to WHO by CIEL) (in German).

Heupt W (1979) [Behaviour of the active ingredients of pesticides in the soil]. (Unpublished report Celamerck No. 111 AA-921-008/009, submitted to WHO by CIEL) (in German).

Heupt W (1983) Determination of the hydrolytic stability of lindane. (Unpublished report Celamerck Ingelheim, submitted to WHO by CIEL).

Hildebrandt G, Jeep H, Hurka H, Stuke T, Heitmann M, & Boiselle C (1986) [Breast milk as a biological indicator: study on the organic halogen burden in breast milk and foodstuffs]. Berlin, Schmidt E. Verlag. (Report 5/86 of the Federal Office for the Environment) (in German).

Hill EF & Camardese MB (1986) Lethal dietary toxicities of environmental contaminants and pesticides to *Coturnix*. Washington, DC, US Department of Interior, Fish and Wildlife Service, p. 89 (Fish and WildlifeTechnical Report No. 2).

Hill EF, Heath RG, Spano JW, & Williams JD (1975) Lethal dietary toxicities of environmental pollutants to birds. Washington, DC, US Department of Interior Fish and Wildlife Service (Special Scientific Report. Wildlife No. 191).

Hofer K (1953) [Investigations on the efficacy of lindane preparations in cow-sheds and on the influence of gamma-HCH on the ripening of Emmental cheese]. University of Munich, Institute of Food Science, Veterinary Faculty (Thesis) (in German).

Holder JW & Stöhrer G (1989) Review of studies relating to lindane carcinogenicity. Washington, DC, US Environmental Protection Agency, Office of Health and Environmental Assessment, Cancer Assessment Group (Unpublished report).

Hoshizaki H, Niki Y, Tajima H, Terada Y, & Kasahara A (1969) A case of leukemia following exposure to insecticide. Acta haematol jpn **32**(14): 672–677.

Hudson RH, Tucker RK, & Haegele MA (1984) Handbook of toxicology of pesticides to wildlife, 2nd ed. Washington DC, US Department of Interior, Fish and Wildlife Service, p. 49 (Resource Publication No. 153).

International Agency for Research on Cancer (1987) Overall evaluations of carcinogenicity: an updating of IARC Monographs Volumes 1 to 42. Lyon, pp. 220–222 (IARC monographs on the evaluation of carcinogenic risks to humans. Suppl 7).

International Atomic Energy Agency (1988) Isotope techniques for studying the fate of persistent pesticides in the tropics. Vienna (IAEA-TELDOC-476).

International Register for Potentially Toxic Chemicals (1989) Data profile on lindane. Geneva, United Nations Environment Programme.

References

Ishidate M, Jr & Odashima S (1977) Chromosome tests with 134 compounds on Chinese hamster cells in vitro—A screening for chemical carcinogens. Mutat Res 48: 337–354.

Ito N, Nagasaki H, Arai M, Makiura S, Sugihara S, & Hirao K (1973a) Histopathologic studies on liver tumorigenesis induced in mice by technical polychlorinated biphenyls and its promoting effect on liver tumours induced by benzenehexachloride. J Natl Cancer Inst 51(5): 1637–1646.

Ito N, Nagasaki H, Arai M, Sugihara S, & Makiura S (1973b) Histologic and ultrastructural studies on the hepatocarcinogenicity of benzenehexachloride in mice. J Natl Cancer Inst 51(3): 817–826.

Ito N, Nagasaki H, Aoe H, Sugihara S, Miyata Y, Arai M, & Shirai T (1975) Brief communication: Development of hepatocellular carcinomas in rats treated with benzenehexachloride. J Natl Cancer Inst 54(3): 801–805.

Itokawa H, Schallah A, Weisgerber I, Klein W, & Korte F (1970) [Contributions to ecological chemistry. XXII. Metabolism and residue bahviour of lindane-^{14}C inhigher plants]. Tetrahedron 26: 763–773 (in German).

Iverson F, Ryan JJ, Lizotte R, & Hierlihy SL (1984) In vivo and in vitro binding of α- and γ-hexachlorocyclohexane to mouse liver macromolecules. Toxicol Lett 20(3): 331–335.

Izmerov NF ed. (1983) Lindane. Moscow, Centre of International Projects, GKNT (IRPTC Scientific reviews of Soviet literature on toxicity and hazards of chemicals No. 40).

Jackson MD, Sheets TJ, & Moffett CL (1974) Persistence and movement of BHC in a watershed. Mount Mitchell State Park, North Carolina, 1967–1972. Pestic Monit J 8(3): 202–208.

Jacobs A, Blangetti M, & Hellmund E (1974) [Storage of chlorinated harmful substances found in Rhine water in the fatty tissue of rats]. Wasser 43: 259–274 (in German).

Jaeger U, Podczek A, Haubenstock A, Pirich K, Donner A, & Hruby K (1984) Acute oral poisoning with lindane-solvent mixtures. Vet Hum Toxicol 26(1): 11–14.

Jagnow G, Haider K, & Ellwardt PC (1977) Anaerobic dechlorination and degradation of hexachlorocyclohexane isomers by anaerobic and facultative anaerobic bacteria. Arch Microbiol 115: 285–292.

Jeanne N (1979) Effets du lindane sur la division, le cycle cellulaire et les biosynthèses de deux algues unicellulaires. Can J Bot 57: 1464–1472.

Jeanne-Levain N (1974) Etude des effets du lindane sur la croissance et le developpement de quelques organismes unicellulaires. Bull Soc Zool Fr 99(1): 105–109.

Jedlicka VI, Hermanska Z, Smida I, & Kouba A (1958) Paramyeloblastic leukaemia appearing simultaneously in two blood cousins after simultaneous contact with gammexane (hexachlorocyclohexane). Acta med scand 161(6): 147–151.

Jenssen D & Ramel C (1980) The micronucleus test as part of a short-term mutagenicity test program for the prediction of carcinogenicity evaluated by 143 agents tested. Mutat Res 75: 191–202.

Johnson RD & Manske DD (1976) Residues in food and feed. Pesticide residues in total diet samples (IX). Pestic Monit J 9(4): 157–169.

Johnson RD, Manske DD, New DH, & Podrebarac DS (1979) Pesticides and other chemical residues in infant and toddler total diet samples, August 1974–July 1975. Pestic Monit J 13(3): 87–98.

Joy RM & Albertson TE (1987a) Factors responsible for increased excitability of dentate gyrus granule cells during exposure to lindane. Neurotoxicology 8(4): 517–528.

Joy RM & Albertson TE (1987b) Interactions of lindane with synaptically mediated inhibition and facilitation in the dentate gyrus. Neurotoxicology 8(4): 529–542.

Joy RM, Stark LG, & Albertson TE (1982) Proconsulvant effects of lindane: enhancement of amygdaloid kindling in the rat. Neurobehav Toxicol Teratol 4: 347–354.

Joy RM, Stark LG, & Albertson TE (1983) Proconvulsant action of lindane compared at two different kindling sites in the rat—amygdala and hippocampus. Neurobehav Toxicol Teratol 5: 461–465.

Joy, RM, Burns VW, & Stark LG (1988) Increases in free intracellular Ca^{++} accompany exposure of neurohybridoma cells to the insecticide lindane. Toxicology 8(1): 44.

Juhnke J & Lüdemann D (1978) [Results of the testing of 200 chemical compounds for acute toxicity to fish in the golden orfe test]. Wasser- Abwasser-Forsch 11(5): 161–164 (in German).

Kahunyo JM, Froslie A, & Maitai CK (1988) Organochlorine pesticide residues in chicken eggs: a survey. J Toxicol Environ Health 24: 543–550.

Kamada T (1971) [Part I. Accumulation of BHC (α-, β-, γ-, and δ-) isomers in rat bodies and excretion into urine following oral administration]. Nippon Eiseigaku Zasshi 26(4): 358-364 (in Japanese).

Kampe W (1980) [Occurrence of hexachlorocyclohexane in the soil]. In: [Hexachlorocyclohexane as a harmful substance in foodstuffs]. Weinheim, Verlag Chemie, pp. 18–23 (Research Report) (in German).

Kampe W & Andre W (1980) [Heavy metals, nitrate and chlorinated hydrocarbons in the vegetable components of total weekly food consumption]. Ernähr Umsch 27(11): 356–364 (in German).

Kaphalia BS & Seth TD (1983) Chlorinated pesticide residues in blood plasma and adipose tissue of normal and exposed human population. Indian J Med Res 77: 245–247.

Karapally JC, Saha JG, & Lee YW (1973) Metabolism of lindane-^{14}C in the rabbit: Ether-soluble urinary metabolites. J Agric Food Chem 21(5): 811–818.

Kathpal TS, Yadav GS, Kushawa KS, & Singh G (1988) Research communication: persistence behavior of HCH in rise soil and its uptake by rice plants. Ecotoxicol Environ Saf 15: 336–338.

Katz M (1961) Acute toxicity of some organic insecticides to three species of salmonids and to the three-spine stickleback. Trans Am Fish Soc 90(3): 264–268.

Kawahara T (1972) [Studies on organochlorine pesticide residues in crops and soils. Report XV. Absorption of BHC by stems and leaves of rice plants]. Noyaku Kensasho Hokohu 12: 31-34 (in Japanese, with English summary).

Kawahara T & Nakamura T (1972) [Studies on the residual organochlorine pesticides in crops and soils. Report XXI. Absorption and translocation of chlorinated hydrocarbons in potted turnips]. Bull Agric Chem Insp Stn 12: 52–56 (Abstract) (in Japanese).

Kawahara T, Konuma S, Wada T, Kureha Y, & Nakamura H (1971) [Absorption and translocation of chlorinated hydrocarbons into crops in the high-cooling region]. Noyaku Kensasho Hokohu 11: 67–72 (in Japanese, with English summary).

References

Kay BD & Elrick DE (1967) Adsorption and movement of lindane in soils. Soil Sci **104**(5): 314–322.

Kensler TW & Trush MA (1984) The role of oxygen radicals in tumour promoters. Environ Mutagen **6**: 593–616.

Kewitz H & Reinert H (1952) [Demonstration of functional changes to the central nervous system through electrically and chemically induced spasma]. Naunyn-Schmiedebergs Arch Exp Pathol Pharmakol **215**: 93–99 (in German).

Khera KS, Whalen C, Trivett G, & Angers G (1979)Teratogenicity studies on pesticidal formulations of dimethoate, diuron, and lindane in rats. Bull Environ Contam Toxicol **22**: 522–529.

Kiraly J, Szentesi I, Ruzicska M, & Czeize A (1979) Chromosome studies in workers producing organophosphate insecticides. Arch Environ Contam Toxicol **8**: 309–319.

Kitamura S, Sumino K, & Hayakawa K (1970) [Decomposition of γ-BHC *in vivo*]. Jpn J Public Health **17**(11): 108–109 (in Japanese).

Klone PR & Kintigh WJ (1988) Lindane technical—fourteen week dust aerosol inhalation study on mice. Bushy Run Research Center (Unpublished report Met Path No. 14014; Laboratory Project No. ID BRRC/51-524).

Koelling K (1978) [Residues in poultry and eggs. Report of the symposium, 28 Mai 1975]. Boppard, Harald Boldt Verlag, pp. 9–159 (in German).

Kohli J, Weisgerber I, Klein W, & Korte F (1976a) Contributions to ecological chemistry: CVII. Fate of lindane-^{14}C in lettuce, endives, and soil under outdoor conditions. J Environ Sci Health **B11**(1): 23–32.

Kohli J, Weisgerber I, & Klein W (1976b) Balance of conversion of [^{14}C]-lindane in lettuce in hydroponic culture. Pestic Biochem Physiol **6**: 91–97.

Kohnen R, Haider K, & Jagnow G (1975) Investigations on the microbial degradation of lindane in submerged and aerated moist soil. Environ Qual Saf **3**: 222–225.

Kolmodin-Hedman B (1974) Exposure to lindane and DDT and its effects on drug metabolism and serum lipoproteins. Stockholm, Departments of Clinical Pharmacology (at Huddinge Hospital) and Pharmacology, Karolinska Institute, Toxicology (Swedish Medical Research Council) and Division of Occupational Medicine, National Board of Occupational Safety and Health (Thesis).

Kolmodin-Hedman B (1984) Pesticides. In: Aitio A, Riihimaki V, & Vainio H, ed. Biological monitoring and surveillance of workers exposed to chemicals. Washington, DC, Hemisphere Publishing Corp., pp. 209–218.

Kolmodin-Hedman B, Alexanderson B, & Sjöqvist F (1971) Effect of exposure to lindane on drug metabolism: Decreased hexobarbital sleeping-times and increased antipyrine disappearance rate in rats. Toxicol Appl Pharmacol **20**(3): 299–307.

Kopf W & Schwoerbel J (1980) [Accumulation studies of the insecticide lindane (γ-HCH, BHC) by aquatic insects of the genus *Sigara* (*Hemiptera, Corixidae*)]. Arch Hydrobiol Suppl **59**(1): 32–42 (in German).

Koransky W, Portig J, & Münch G (1963) [Absorption, distribution and excretion of α- and γ-hexachlorocyclohexane]. Naunyn-Schmiedebergs Arch Exp Pathol Pharmakol **244**: 564–575 (in German).

Koransky W, Portig J, Vohland HW, & Klempau I (1964) [Elimination of α- and γ-hexachlorocyclohexane and the way it is influenced by liver microsomal enzymes]. Naunyn-Schmiedebergs Arch Exp Pathol Pharmakol **247**: 49–60 (in German).

Korn S & Earnest R (1974) Acute toxicity of twenty insecticides to striped bass, *Morone saxatilis*. California Fish Game **60**(3): 128–131.

Korte F (1980) [Occurrence of HCH-hexachlorcyclohexane as a harmful substance in foods. Papers from two symposia of the Senate Committee on the Testing of Residues in Foods, held on 28–29 November 1979 and 6 March 1980]. Weinheim, Verlag Chemie, pp. 46–64 (in German).

Kotzias D, Nitz S, & Korte F (1981) [Light-induced total breakdown of organic molecules adsorbed on silica gel]. Chemosphere **10**(4): 415–422 (in German).

Kramer MS, Hutchinson TA, Rudnick SA, Leventhal JM, & Feinstein AR (1980) Operational criteria for adverse drug reactions in evaluating suspected toxicity of a popular scabicide. Clin Pharmacol Ther **27**: 149–154.

Krauthacker B, Alebic-Kolbah T, Kralj M, Tjalcevic B, & Reiner E (1980) Organochlorine pesticides in blood serum of the general Yugoslav population and in occupationally exposed workers. Int Arch Occup Environ Health **45**: 217–220.

Krishnakumari MK (1977) Sensitivity of the alga *Scenedesmus acutus* to some pesticides. Life Sci **20**: 1525–1532.

Kujawa M, Härtig M, Macholz RM, & Engst R (1976) [The degradation of ^{14}C-lindane by a mould culture]. Nahrung **20**(2): 181–183 (in German).

Kujawa M, Engst R, & Macholz R (1977) On the metabolism of lindane. In: Proceedings of International Symposium on Industrial Toxicology, Environmental Pollution and Human Health, pp. 661–672.

Kumar, A. & Dwivedi PP (1988) Relative induction of molecular forms of cytochrome P-450 in γ-hexachlorocyclohexane exposed rat liver microsomes. Arch Toxicol **62**: 479–481.

Kurihara N & Nakajima M (1974) Studies on BHC isomers and related compounds. VIII. Urinary metabolites produced from γ- and β-BHC in the mouse: Chlorophenol conjugates. Pestic Biochem Physiol **4**: 220–231.

Kurihara N, Uchido M, Fujita T, & Nakajima M (1973) Studies on BHC isomers and related compounds. V. Some physicochemical properties of BHC-isomers(1). Pestic Biochem Physiol **2**: 383–390.

Kurihara N, Tanaka K, & Nakajima M (1979) Mercapturic acid formation from lindane in rats. Pestic Biochem Physiol **10**: 137–150.

Kurihara N, Suzuki T, & Nakajima M (1980) Deuterium isotope effects on the formation of mercapturic acids from lindane in rats. Pestic Biochem Physiol **14**: 41–49.

Kurt TL, Dost R, Gilliland M, Reed G, & Petty C (1986) Accidental Kwell (lindane) ingestions. Vet Hum Toxicol **28**(6): 569–571. (Published erratum appears in Vet Hum Toxicol **29**(1): 24, 1987.)

Landesamt für Wasser und Abfall (1988) [Water quality report 1987]. Dusseldorf, North Rhine-Westphalia Office for Water and Refuse (in German).

References

Lange P (1965) [Studies on the convulsive process and the effect of convulsion-inhibiting substances thereon during progressive infusion of convulsive poisons in mice]. Acta Biol Med Ger 15: 109–119 (in German).

Lange M, Nitzsche K, & Zesch A (1981) Percutaneous absorption of lindane in healthy volunteers and scabies patients. Dependency of penetration kinetics in serum upon frequency of application, time and mode of washing. Arch Dermatol Res 271: 387–399.

Laug EP (1948) Tissue distribution of a toxicant following oral ingestion of the gamma isomer of benzene hexachloride by rats. J Pharmacol Exp Ther 93: 277–281.

Laugel P (1981) [The residue situation in France and ways of overcoming it]. Lebensmittelchem Gerichtl Chem 35: 29–32 (in German).

Lehman AJ (1952) Chemicals in foods: A report to the association of Food and Drug officials on current developments. Part II. Pesticides. Section III. Subacute and chronic toxicity. Q Bull Assoc Food Drug Off US 16(2): 47–53.

Lehman AJ (1965) Summaries of pesticide toxicity. Part I. Chlorinated organic compounds. Topeka, Kansas, Associations with Food and Drug Officials of the United States, pp. 27–29.

Lichtenstein EP & Polivka JB (1959) Persistence of some chlorinated hydrocarbon insecticides in turf soils. J Econ Entomol 52(2): 289–293.

Lichtenstein EP & Schulz KR (1958a) Persistence of some chlorinated hydrocarbon insecticides as influenced by soil types, rate of application and temperature. J Econ Entomol 52(1): 124–131.

Lichtenstein EP & Schulz KR (1958b) Breakdown of lindane and aldrin in soils. J Econ Entomol 52(1): 118–124.

Lichtenstein EP & Schulz KR (1970) Volatilization of insecticides from various substrates. J Agric Food Chem 18: 814.

Lichtenstein EP, de Peu LJ, Eshbaugh EL, & Sleesman JP (1960) Persistence of DDT, aldrin, and lindane in some Midwestern soils. J Econ Entomol 53(1): 136–142.

Lisi P, Caraffini S, & Assalve D (1986) A test series for pesticide dermatitis. Contact Derm 15: 266–269.

Lisi P, Caraffini S, & Assalve D (1987) Irritation and sensitization potential of pesticides. Contact Derm 17: 212–218.

Lowden GF, Saunders CL, & Edwards RW (1969) Organochlorine insecticides in water (Part II). J Water Treat Exam 18: 275–287.

Lüdemann D & Neumann H (1960a) [Tests on the acute and toxic effects of modern contact insecticides on one-year-old carp (*Cyprinus carpio* L.)]. Z Angew Zool 47: 11–33 (in German).

Lüdemann D & Neumann H (1960b) [Tests on the acute and toxic effects of modern contact insecticides on freshwater animals]. Z. Angew Zool 47: 303–321 (in German).

Macek KJ & McAllister WA (1970) Insecticide susceptibility of some common fish family representatives. Trans Am Fish Soc 99: 20–27.

Macek KJ, Hutchinson C, & Cope OB (1969) The effects of temperature on the susceptibility of bluegills and rainbow trout to selected pesticides. Bull Environ Contam Toxicol 4(3): 174–183.

Macek KJ, Buxton KS, Derr SK, Dean JW, & Sauter S (1976) Chronic toxicity of lindane to selected aquatic invertebrates and fish. Washington, DC, US Environmental Protection Agency (Ecological Research Series, EPA-600/3-76-046).

Macholz RM, Knoll R, Lewerenz HJ, Plass R, & Schulze J (1983) [Metabolism of gamma-hexachlorocyclohexane (HCH) in germ-free and conventional rats]. Zentralbl Pharmakol 122(2): 221 (in German).

MacKay D & Walkoff AW (1973) Rate of evaporation of low-solubility contaminants from water bodies to atmosphere. Environ Sci Technol 7: 611.

MacRae IC, Raghu K, & Castro TF (1967) Persistence and biodegradation of four common isomers of benzenehexachloride in submerged soils. J Agric Food Chem 15(5): 911–914.

MacRae JC, Yamaya Y, & Yoshida T (1984) Persistence of HCH-isomers in soil suspensions. Soil Biol Biochem 16: 285–286.

Marcelle C & Thome JP (1983) Acute toxicity and bioaccumulation of lindane in gudgeon, *Gobio gobio* (L.). Bull Environ Contam Toxicol 31: 453–458.

Marcelle C & Thome JP (1984) Relative importance of dietary and environmental sources of lindane in fish. Bull Environ Contam Toxicol 33: 423–429.

Marchal-Segault D & Ramade F (1981) The effects of lindane, an insecticide, on hatching and post-embryonic development of *Xenopus laevis* (Daudin). Anuran amphibian. Environ Res 24: 250–258.

Martin RJ & Duggan RE (1968) Pesticide residues in total diet samples (III). Pestic Monit J 1(4): 1–20.

Martin DB & Hartmann WA (1985) Organochlorine pesticides and polychlorinated biphenyls in sediment and fish from wetlands in the North-Central United States. J Assoc Off Anal Chem 69(4): 712–717.

Mathur SP & Saha JG (1975) Microbial degradation of lindane-^{14}C in a flooded sandy loam soil. Soil Sci 120(4): 301–307.

Mathur SP & Saha JG (1977) Degradation of lindane-^{14}C in a mineral soil and in an organic soil. Bull Environ Contam Toxicol 17(4): 424–430.

Matsumura F (1985) Toxicology of insecticides, 2nd ed. New York, Plenum Press, pp. 128–143.

Matsumura F & Benezet HJ (1973) Studies on the bioaccumulation and microbial degradation of 2,3,7,8,-tetrachlorodibenzo-p-dioxin. Environ Health Perspect 5: 253–258.

Matsumura F & Tanaka K (1984) Molecular basis of neuroexcitability actions of cyclodiene-type insecticides. In: Narahashi T, ed., Cellular and molecular neurotoxicology. New York, Raven Press, pp. 225–240.

Matsumura F, Benezet HJ, & Patil KC (1976) Factors affecting microbial metabolism of γ-BHC. J Pestic Sci 1: 3–8.

McLeay DJ (1976) A rapid method for measuring the acute toxicity of pulpmill effluents and other toxicants to salmonid fish at ambient room temperature. J Fish Res Board Canada 33: 1303–1311.

References

McNamara B & Krop S (1948) The treatment of acute poisoning produced by gamma hexachlorocyclohexane. J Pharmacol **92**: 147–152.

McParland PJ, McCracken RM, O'Hare MB, & Raven AM (1973) Benzenehexachloride poisoning in cattle. Vet Rec **93**: 369–371.

Meemken HA, Habersaat K, & Groebel W (1982) [Burden of breast milk with chlorinated hydrocarbons from creams containing anhydrous lanolin]. Lebensmittelchem Gerichtl Chem **36**: 51-53 (in German).

Mendeloff AI & Smith DE (1955) Exposure to insecticides, bone marrow failure, gastrointestinal bleeding, and uncontrollable infections. Am J Med **19**(8): 274–284.

Mes J, Davies DJ, & Turton D (1982) Polychlorinated biphenyl and other chlorinated hydrocarbon residues in adipose tissue of Canadians. Bull Environ Contam Toxicol **28**: 97–104.

Mes J, Davies DJ, Turton D, & Sun WF (1986) Levels and trends of chlorinated hydrocarbon contaminants in the breast milk of Canadian women. Food Addit Contam 3(4): 313–322.

Mestres R (1974) Report of a group of experts on the content of organohalogen compounds detected between 1968 and 1972 in water, air and foodstuffs and the methods of analysis used in the nine Member States of the European Community. Presented at the European Colloquium on problems raised by the contamination of man and his environment by persistent pesticides and organo-halogenated compounds, Luxembourg, 14–16 May 1974. Luxembourg, Commission of the European Communities (Paper No. 1).

Milby TH & Samuels AJ (1971) Human exposure to lindane—comparison of an exposed and unexposed population. J Occup Med **13**(5): 256–258.

Milby TH, Samuels AJ, & Ottoboni F (1968) Human exposure to lindane; blood lindane levels as a function of exposure. J Occup Med **10**(10): 584–587.

Mills AC & Biggar JW (1969) Solubility–temperature effect on the adsorption of gamma- and beta-BHC from aqueous and hexane solutions by soil materials. Soil Sci Soc Am Proc **33**: 210–216.

Morgan DP, Stockdale EM, Roberts RJ, & Walter AW (1980) Anemia associated with exposure to lindane. Arch Environ Health **35**(5): 307–310.

Mosha RD, Gyrd-Hansen N, & Kraul I (1986) Distribution and elimination of lindane in goats. Bull Environ Contam Toxicol **36**: 518–522.

Mottram DS, Psomas IE, & Patterson RLS (1983) Chlorinated residues in the adipose tissue of pigs treated with γ-hexachlorocyclohexane. J Sci Food Agric **34**: 378–387.

Mouvet C (1985) Monitoring of polychlorinated biphenyls (PCBs) and hexachlorocyclohexanes (HCH) in fresh water using the aquatic moss *Cinclidotus danubicus*. Sci Total Environ **44**: 253–267.

Moza P, Mostafa I, & Klein W (1974) [ntributions to ecological chemistry. LXXXIX. Exploratory investigations of the metabolism of γ-pentachlorcyclohex-1-ene in higher plants grown hydroponically]. Chemosphere **6**: 255–258 (in German).

Muacevic G (1966) [Oral toxicity in male rats]. Ingelheim am Rhein, Boehringer & Son (Unpublished report submitted to WHO by CIEL) (in German).

Muacevic G (1970) [LD$_{50}$ determination in female rats]. Ingelheim am Rhein, Boehringer & Son (Unpublished report submitted to WHO by CIEL) (in German).

Muacevic G (1971a) [LD$_{50}$ determination in male rats]. Ingelheim am Rhein, Boehringer & Son (Unpublished report submitted to WHO by CIEL) (in German).

Muacevic G (1971b) [LD$_{50}$ determination in the rat]. Ingelheim am Rhein, Boehringer & Son (Unpublished report submitted to WHO by CIEL) (in German).

Müller D, Klepel H, Macholz RM, Lewerenz HJ, & Engst R (1981) Electroneurophysiological studies on neurotoxic effects of hexachlorocyclohexane isomers and gamma-pentachlorocyclohexene. Bull Environ Contam Toxicol 27: 704–706.

Muralidhara MK, Krishnakumari MK, & Majumder SK (1979) Effects of carriers on the oral toxicity of lindane (γ-BHC) to albino rats. J Food Sci Technol 16: 105–107.

Nair GA (1981) Toxic effects of certain biocides on a freshwater mite, *Hydrachna trilobata* Viets (Arachnida; Hydrachnoidea; Hydrachnidae). J Environ Biol 2(2): 91–96.

Nash RG (1983) Determining environmental fate of pesticides with microagroecosystems. Residue Rev 85: 199–215.

Nash RG, Harris WG, Ensor PD, & Woolson EA (1973) Comparative extraction of chlorinated hydrocarbon insecticides from soils 20 years after treatment. J Assoc Off Anal Chem 56(3): 728–732.

Newland LW, Chesters G, & Lee GB (1969) Degradation of γ-BHC in simulated lake impoundments as affected by aeration. J Water Pollut Control Fed 41: R174–R188.

Niessen KH, Ramolla J, Binder M, Brügmann G, & Hofmann U (1984) Chlorinated hydrocarbons in adipose tissue of infants and toddlers: inventory and studies on their association with intake of mothers' milk. Eur J Pediatr 142: 238–243.

Nigam SK, Karnik AB, Majumder SK, Visweswarriah K, Suryanarayana Raju G, Muktha Bai K, Lakkad BC, Thakore KN, & Chatterjee BB (1986) Serum hexachlorocyclohexane residues in workers engaged at a HCH manufacturing plant. Int Arch Occup Environ Health 57: 315–320.

Nitsche K, Lange M, Bauer E, & Zesch A (1985) Quantitative distribution of locally applied lindane in human skin and subcutaneous fat *in vitro*. Dermatosen 32(5): 161–165.

Noren K (1983) Levels of organochlorine contaminants in human milk in relation to the dietary habits of the mothers. Acta Paediatr Scand 72: 811–816.

Norstrom RJ, Simon M, Muir DCG, & Schweinsburg RE (1988) Organochlorine contaminants in Arctic marine food chains: identification, geographical distribution and temporal trends in polar bears. Environ Sci Technol 22: 1063–1071.

Nurmatov RS (1965) [Effect of trichlormetaphos-3 and lindane on animals]. Veterinarija 44(8): 85–87 (in Russian).

Ocker HD (1983) [Contaminants in cereal grains—state of knowledge and list of deficiencies]. Getreide-Mehl-Brot 37: 3–7 (in German).

Oehme M & Mano S (1984) The long-range transport of organic pollutants in the Arctic. Fresenius Z Anal Chem 319: 141–146.

References

Oehme M & Stray H (1982) Quantitative determination of ultra-traces of chlorinated compounds in high-volume air samples from the Arctic using polyurethane foam as collecting medium. Fresenius Z Anal Chem 311: 655–673.

Oesch F (1980) Bacterial mutagenicity tests of lindane with mouse liver preparations as metabolizing systems. University of Mainz (Unpublished report Celamerck No. 111AA-457-006, submitted to WHO by CIEL).

Oesch F & Glatt HR (1984) Mammalian cell (V79) mutagenicity test on lindane. University of Mainz (Unpublished report Celamerck No. 111AC-457- 019/SP 540-VT21, submitted to WHO by CIEL).

Oesch F, Friedberg T, Herbst M, Paul W, Wilhelm N, & Bentley P (1982) Effects of lindane treatment on drug metabolizing enzymes and liver weight of CF1 mice in which it evoked hepatomas and in non-susceptible rodents. Chem-biol Interactions 40: 1–14.

Ohisa N & Yamaguchi M (1978a) Degradation of gamma-BHC in flooded soils enriched with peptone. Agric Biol Chem 42(11): 1983–1987.

Ohisa N & Yamaguchi M (1978b) Gamma BHC degradation accompanied by the growth of *Clostridium rectum* isolated from paddy field soil. Agric Biol Chem 42(10): 1819–1823.

Ohisa N & Yamaguchi M (1979) *Clostridium* species and γ-BHC degradation in paddy soil. Soil Biol Biochem 11: 645–649.

Ohisa N, Yamaguchi M, & Kurihara N (1980) Lindane degradation by cell free extracts of *Clostridium rectum*. Arch Microbiol 125: 221–225.

Ohisa N, Kurihara N, & Nakajima M (1982) ATP synthesis associated with the conversion of hexa-chlorocyclohexane related compounds. Arch Microbiol 131: 330–333.

Ohly A (1973) Toxicology of lindane. In: Ulmann, E. ed., Lindane, Vol. I, Supplement 1974. Brussels, CIEL, p. 27.

Oldiges H, Takenaka S, & Hochrainer D (1980) [Inhalation test with lindane (γ-hexachlorocyclohexane) to determine the LC_{50}]. (Unpublished report Celamerck No. 111AA-423-001 submitted to WHO by CIEL) (in German).

Oldiges H, Hertel R, Kördel W, Hochrainer D, & Mohr U (1983) [90-Day inhalation with lindane]. Schmallenberg, Fraunhofer Institut für Toxikologie und Aerosolforschung (Unpublished report Celamerck No. 111AC-435-004, submitted to WHO by CIEL) (in German).

Oloffs PC, Albright LJ, Szeto SY, & Lau J (1973) Factors affecting the behaviour of five chlorinated hydrocarbons in two natural waters and their sediments. J Fish Res Board Canada 30(11): 1619–1623.

Oshiba K (1972) Experimental studies on the fate of β- and γ-BHC *in vivo* following daily administration. Osaka Shiritsu Daigaku Igaku Zasshi 21(1–3): 1–19.

Paasivirta J, Palm H, Paukka R, Akhabuhaya J, & Lodenius M (1988) Chlorinated insecticide residues in Tanzanian environment, Tanzadrin. Chemosphere 17(10): 2055–2062.

Palmer CM & Maloney TE (1955) Preliminary screening for potential algicides. Ohio J Sci 55(1): 1–8.

Palmer L & Kolmodin-Hedman B (1972) Improved quantitative gas chromatographic method for the analysis of small amounts of chlorinated hydrocarbon pesticides in human plasma. J Chromatogr 74: 21–30.

Palmer AK, Cozen DD, Spicer EJF, & Worden AN (1978a) Effects of lindane upon reproductive function in a 3-generation study in rats. Toxicology 10: 45–54.

Palmer AK, Bottomley AM, Worden AN, Frohberg H, & Bauer A (1978b) Effect of lindane on pregnancy in the rabbit and rat. Toxicology 9: 239–247.

Pandy RN, Zaidi SIM, & Kidawi AM (1985) Effect of pesticides on frog skeletal muscle sarcolemmal enzymes. J Environ Biol 6(1): 7–9.

Panwar RS, Gupta RA, Joshi HC, & Kapoor D (1982)Toxicity of some chlorinated hydrocarbons and organophosphorus insecticides to gastropod, *Viviparus bengalensis* Swainson. J Environ Biol 3(1): 31–36.

Pastor A, Hernandez F, Medina J, Melero R, Lopez FJ, & Conesa M (1988) Organochlorine pesticides in marine organisms from the Castellon and Valencia coasts of Spain. Mar Pollut Bull 19(5): 235–238.

Paul W, Knappen F, & Stötzer H (1980) [Testing for acute toxicity after oral administration in an oily solution to Chbi: NMRI (SPF) mice]. Ingelheim am Rhein, Boehringer & Son (Unpublished report submitted to WHO by CIEL) (in German).

Petring OU, Adelhoj B, & Jorgensen FR (1986) Acute poisoning after oral intake of lindane. Ugeskr Laeg 148(50): 3377.

Platford RF (1981) The environmental significance of surface films. II. Enhanced partitioning of lindane in thin films of octanol on the surface of water. Chemosphere 10(7): 719–722.

Podlejski J & Dervieux AM (1978) Persistance de quatre produits phytosanitaires dans les eaux de rivières en Camargue: caractérisation de leurs effets sur l'écosystème. Trav Soc Pharm Montpellier 38(2): 153–164.

Polishuk ZW, Wassermann M, Wassermann D, Groner Y, Lazarovici S, & Tomatis L (1970) Effects of pregnancy on storage of organochlorine insecticides. Arch Environ Health 20: 215–217.

Portig J, Kraus P, Stein K, Koransky W, Noack G, Gross B, & Sodomann S (1979) Glutathione conjugate formation from hexachlorocyclohexane and pentachlorocyclohexene by rat liver *in vitro*. Xenobiotica 9: 353–378.

Portmann JE (1970) Results of acute toxicity test with marine organisms, using a standard method. In: Ruivo, M. ed., Marine pollution and sea life. London, Fishing News (Books) Ltd, pp. 212–217.

Portmann JE (1979) Evaluation of the impact on the aquatic environment of hexachlorocyclohexane (HCH-isomers). Brussels, Commission of the European Communities (Contract No. U/78/180-SEC (78)792).

Probst GS, McMahon RE, Hill LE, Thompson CZ, Epp JK, & Neal SB (1981) Chemically-induced unscheduled DNA synthesis in primary rat hepatocyte cultures: A comparison with bacterial mutagenicity using 218 compounds. Environ Mutagen 3: 11–32.

References

Publicover SJ & Duncan CJ (1979) The action of lindane in accelerating the spontaneous release of neurotransmitter at the frog neuromuscular junction. Naunyn-Schmiedebergs Arch Pharmacol **381**: 179–182.

Publicover SJ, Duncan CJ, & Smith JL (1979) The action of lindane in causing ultrastructural damage in frog skeletal muscle. Comp Biochem Physiol **C64**: 237–241.

Racey PA & Swift SM (1986) The residual effects of remedial timber treatments on bats. Biol Conserv **35**: 205–214.

Ramamoorthy S (1985) Competition of fate processes in the bioconcentration of lindane. Bull Environ Contam Toxicol **34**: 349–358.

Randall WF, Dennis WH, & Warner MC (1979) Acute toxicity of dechlorinated DDT, chlordane, and lindane to bluegill (*Lepomis macrochirus*) and *Daphnia magna*. Bull Environ Contam Toxicol **21**: 849–854.

Rao MB (1981) Effect of γ-hexachloran and Sevin on the survival of the black sea mussel, *Mytilus galloprovincialis* Lam. Hydrobiologia **78**: 33–37.

Rao PSC & Davidson JM (1982) Retention and transformation of selected pesticides and phosphorus in soil-water system: a critical review. Springfield, Virginia, US Department of Commerce, National Technical Information Services (EPA Report No. 600/3-82-060, PB 92- 256884).

Rao SM & Reddy JK (1987) Peroxisome proliferation and hepatocarcinogenesis. Carcinogenesis **8**(5): 631–636.

Rappe C, Nygren M, Lindström G, Buser HR, Blaser O, & Wüthrich C (1987) Polychlorinated dibenzofurans and dibenzo-p-dioxins and other chlorinated contaminants in cow milk from various locations in Switzerland. Environ Sci Technol **21**: 964–970.

Raw GR, ed. (1970) CIPAC Handbook. Vol. 1. Analysis of technical and formulated pesticides. Collaborative International Pesticides Analytical Council Ltd, New York, Academic Press, pp. 71–78, 129–131.

Ray RC (1983) Toxicity of the pesticides hexachlorocyclohexane and Benomyl to nitrifying bacteria in flooded autoclaved soil and in culture media. Environ Pollut **A32**: 147–155.

Reifenrath WG, Chellquist EM, Shipwash EA, Jederberg WW, & Krueger GG (1984) Percutaneous penetration in the hairless dog, weanling pig and grafted athymic nude mouse: evaluation of models for predicting skin penetration in man. Br J Dermatol **27** (Suppl. III): 123–135.

Reiner E, Krauthacker B, Stipcevic M, & Stefanac Z (1977) Blood levels of chlorinated hydrocarbon residues in the population of a continental town in Croatia (Yugoslavia). Pestic Monit J **11**: 54–55.

Reno FE (1976a) Teratology study in rats: Lindane (gamma benzene hexachloride, USP). Madison, Wisconsin, Hazelton Laboratories America Inc. (Unpublished report submitted to WHO by CIEL).

Reno FE (1976b) Teratology study in rabbits: Lindane (gamma benzene hexachloride, USP). Madison, Wisconsin, Hazelton Laboratories America Inc. (Unpublished report submitted to WHO by CIEL).

Rhône-Poulenc Agrochimie (1986) Lindane. Information kit. Lyon.

Riemschneider R (1949) [A contribution to the toxicity of contact insecticides]. Anz Schädlingskd **22**(1): 1–3 (in German).

Rivett KF, Chesterman H, Kellett DN, Newman AJ, & Worden AN (1978) Effects of feeding lindane to dogs for periods of up to 2 years. Toxicology **9**: 273–289.

Rocchi P, Perocco P, Alberghini W, Fini A, & Prodi G (1980) Effect of pesticides on scheduled and unscheduled DNA synthesis of rat thymocytes and human lymphocytes. Arch Toxicol **45**: 101–108.

Röhrborn G (1976) Cytogenetic analysis of bone marrow of Chinese hamster (*Cricetulus grisens*) after sub-acute treatment with lindane. (Unpublished report Celamerck No. 111AA-457-008, submitted to WHO by CIEL).

Röhrborn G (1977a) Statement on the potential mutagenicity of lindane (γ-hexachlorocyclohexane). (Unpublished report Celamerck No. 111AA-457-018, submitted to WHO by CIEL).

Röhrborn G (1977b) Dominant lethal test after treatment of male rats with lindane. (Unpublished report Celamerck No. 111AA-457-004, submitted to WHO by CIEL).

Rosenberg LE, Rudd RL, Devaul J, Lundberg DE, Recca JM, & Smith MB (1953) The effects of economic poisons on wildlife. Quarterly progress report of the Federal Aid in Wildlife Restoration Act pp. 2, 7, 11–12, 19, 43–45 (Project No. W-45R-1) (Unpublished report).

Sagelsdorff P, Lutz WK, & Schlatter C (1983) The relevance of covalent binding to mouse liver DNA to the carcinogenic action of hexachlorocyclohexane isomers. Carcinogenesis **4**(10): 1267–1273.

Samuels AJ & MilbyTH (1971) Human exposure to lindane. Clinical, haematological, and biochemical effects. J Occup Med **13**(3): 147–151.

San Antonio JP (1959) Demonstration of lindane and a lindane metabolite in plants by paper chromatography. J Agric Food Chem **7**(5): 322–325.

Sanders HO (1969) Toxicity of pesticides to the crustacean *Gammarus lacustrus*. Washington, DC, US Department of Interior, Bureau of Sport Fisheries and Wildlife (Technical Paper No. 25), pp. 2–18.

Sanders HO (1970) Pesticide toxicities to tadpoles of the western chorus frog *Pseudacris triseriata* and Fowler's toad *Bufo woodhousii fowleri*. Copeia **2**: 246–251.

Sanders HO & Cope OB (1966) Toxicities of several pesticides to two species of *Cladocerans*. Trans Am Fish Soc **95**: 165–169.

Sanders HO & Cope OB (1968) The relative toxicities of several pesticides to naiads of three species of stoneflies. Limnolog Oceanogr **13**: 112–117.

Sanfeliu C, Camon L, Martinez E, Sola C, Artigas F, & Rodriguez-Farre E (1988) Regional distribution of lindane in rat brain. Toxicology **49**: 189–196.

Saxena MC, Siddique MKJ, Bhargava AK, Krishna Murti CR, & Kutty D (1981) Placental transfer of pesticides in humans. Arch Toxicol **48**: 127–134.

Schafer EW (1972) The acute oral toxicity of 369 pesticidal, pharmaceutical and other chemicals to wild birds. Toxicol Appl Pharmacol **21**: 315–330.

References

Schimmel SC, Patrick JM, Jr, & Forester J (1977) Toxicity and bioconcentration of BHC and lindane in selected animals. Arch Environ Contam Toxicol 6: 355–363.

Schmiedeberg J & Wasserberger HJ (1953) [Acute toxicity of hexachlorocyclohexane for man]. Anz Schädlingskd 26(9): 129–133 (in German).

Schmitt F (1956) [Experimental investigations on the behaviour of hexachlorocyclohexane and aldrin in the soil, with special reference to duration of activity, effects on small animal fauna, and impairment of the taste of geocarpic fruits]. Hohenheim, Agricultural College (Thesis) (in German).

Schneeweis JC, Greichus TA, & Linder RL (1974) Organochlorine pesticide residue levels in North American timber wolves—1969–71. Pestic Monit J 8(2): 142–143.

Schnell U (1965) [Investigations on the chronic toxicity of lindane for pigs, with special reference to investigations for residues in fat and liver]. Berlin, Humboldt University (Thesis) (in German).

Schröter C, Parzefall W, Schröter H, & Schulte-Hermann R (1987) Dose–response studies on the effects of α-, β- and γ-hexachlorocyclohexane on putative preneoplastic foci, monooxygenases, and growth in rat liver. Cancer Res 47: 80–88.

Schüttmann W (1972) Clinical observations on the long-term toxicity of organochlorine pesticides. Z Gesamte Hyg 17: 12–18 (in German).

Schwabe U & Wendling I (1967) [Stimulation of drug metabolism by low doses of DDT and other chlorinated hydrocarbon insecticides]. Arzneimittelforschung 17(5): 614–618 (in German).

Seidler H, Macholz RM, Härtig M, Kujawa M, & Engst R (1975) [Studies on the metabolism of certain insecticides and fingicides in the rat. Part IV. Distribution, degradation and excretion of ^{14}C-labelled lindane]. Nahrung 19(5/6): 473–482 (in German).

Selby LA, Newell KW, Hauser GA, & Junker G (1969) Comparison of chlorinated hydrocarbon pesticides in maternal blood and placental tissues. Environ Res 2: 247–255.

Seuge J & Bluzat R (1979a) Toxicité chronique du carbaryl et du lindane chez le mollusque d'eau douce Lymnea stagnalis L. Water Res 13(3): 285–293.

Seuge J & Bluzat R (1979b) Etude de la toxicité chronique de deux insecticides (carbaryl et lindane) à la génération F1 de Lymnea stagnalis L. (mollusque gastéropode pulmone). II. Conséquences sur le potentiel reproducteur. Hydrobiologia 66(1): 25–31.

Seuge J & Bluzat R (1982) Influence d'une intoxication par le lindane en fonction de la dureté de l'eau chez Lymnea stagnalis (Mollusca pulmonata). Malacologia 22(1–2): 15–18.

Sharom MS, Miles JRW, Harris CR, & McEwen FL (1980) Persistence of 12 insecticides in water. Water Res 14: 1089–1093.

Shearer RC, Letey J, Farmer WJ, & Klute A (1973) Lindane diffusion in soil. Soil Sci Soc Am Proc 37: 189–193.

Shiota Y & Kanda S (1972) [Fate of benzene hexachloride in paddy soils]. Aichi-Ken Nogyo Sogo Shikenjho, Kenkyu Hokoku 1972A/4: 128–133 (in Japanese).

Shirasu Y, Moriya M, Kato K, Furuhashi A, & Kada T (1976) Mutagenicity screening of pesticides in the microbial system. Mutat Res 40: 19–30.

Siddaramappa R & Sethunathan N (1975) Persistence of gamma-BHC and beta-BHC in Indian rice soils under flooded conditions. Pestic Sci 6: 395–403.

Siddique MKJ, Saxena MC, Mishra UK, Krishnamurti CR, & Nag D (1981) Long-term occupational exposure to DDT. Int Arch Occup Environ Health 48: 301–308.

Sierra M & Santiago D (1987) Organochlorine pesticide levels in barn owls collected in Leon, Spain. Bull Environ Contam Toxicol 38: 261–265.

Sina JF, Bean CL, Dysart GR, Taylor VI, & Bradley MO (1983) Evaluation of the alkaline elution/rat hepatocyte assay as a predictor of carcinogenic/mutagenic potential. Mutat Res 113: 357–391.

Sinha RRP & Sinha SP (1983) Induction of dominant lethal mutations in *Drosophila melanogaster* by three pesticides: Gammexane, Sevin, and Folidol. Comp Physiol Ecol 8(2): 87–89.

Skaare JU, Tuveng JM, & Sande HA (1988) Organochlorine pesticides and polychlorinated biphenyls in maternal adipose tissue, blood, milk, and cord blood from mothers and their infants living in Norway. Arch Environ Toxicol 17: 55–63.

Skaftason JF & Johannesson T (1979) Organochlorine compounds (DDT, hexachlorocylohexane, hexachlorobenzene) in Icelandic animal body fat and butter fat: Local and global sources of contamination. Letter to the Editor. Acta Pharmacol Toxicol 44: 156–157.

Slade RE (1945) The γ-isomer of hexachlorocyclohexane (gammexane). Chem Ind 65: 314-319.

Slooff W & Matthijsen AJCM (1988) Integrated criteria document: hexachlorocyclohexanes. Bilthoven, National Institute of Public Health and Environmental Protection (Report No. 758473011).

Srinivasan K & Radhakrishnamurty R (1988) Biochemical changes produced by β- and γ-hexachlorocyclohexane isomers in albino rats. J Environ Sci Health B23(4): 367–386.

Srinivasan K, Ramesh HP, & Radhakrishnamurty R (1984) Renal tubular dysfunction caused by dietary hexachlorocyclohexane (HCH) isomers. J Environ Sci Health B19(4/5): 453–466.

Srinivasan K, Ramesh HP, & Radhakrishnamurty R (1988) Changes induced by hexachlorocyclohexane isomers in rat livers and testes. Bull Environ Contam Toxicol 41: 531–539.

Starr HG & Clifford NJ (1972) Acute lindane intoxication. A case study. Arch Environ Health 25: 374–375.

Starr RI & Johnson RE (1968) Laboratory method for determining the rate of volatilization of insecticides from plants. J Agric Food Chem 16(3): 411–414.

Steering Group on Food Surveillance (1982) Report of the Working Party on Pesticide Residues (1977–1981). London, Her Majesty's Stationery Office (Food Surveillance Paper No. 9).

Steering Group on Food Surveillance (1986) Report of the Working Party on Pesticide Residues (1982–1985). London, Her Majesty's Stationery Office (Food Surveillance Paper No. 16).

Steering Group on Food Surveillance (1989) Report of the Working Party on Pesticide Residues (1985–1988). London, Her Majesty's Stationery Office (Food Surveillance Paper No. 25).

Stein K, Portig J, & Koransky W (1977) Oxidative transformation of hexachlorocyclohexane in rats and with rat liver microsomes. Naunyn-Schmiedebergs Arch Pharmacol 298: 115–128.

References

Stein K, Portig J, Fuhrmann H, Koransky W, & Noack G (1980) Steric factors in the pharmacokinetics of lindane and α-hexachlorocyclohexane in rats. Xenobiotica 10(1): 65–77.

Steinwandter H (1976) [Lindane metabolism in plants. II. Formation of α-HCH]. Chemosphere 4: 221–225 (in German).

Stewart DKR & Chisholm D (1971) Long-term persistence of BHC, DDT, and chlordane in a sandy loam soil. Can J Soil Sci 51: 379–383.

Stewart DKR & Fox CJS (1971) Persistence of organochlorine insecticides and their metabolites in Nova Scotian soils. J Econ Entomol 64(2): 367–371.

Stieglitz R, Stobbe H, & Schüttmann W (1967) [Bone marrow damage following exposure to the insecticide gamma-hexachlorocyclohexane (lindane)]. Acta Haematol 38: 337–350 (in German).

Stöckigt J (1976) Investigations of the metabolism of γ-HCH in plant tissue. (Unpublished report Celamerck No. 111AA-641-004, submitted to WHO by CIEL).

Stöckigt J & Ries B (1976) Catabolism of γ-hexachlorocyclohexane (lindane) by plants and cell cultures, a comparison. Bochum, Chair of Plant Physiology, Ruhr University (Unpublished report Celamerck No. 111AA-641-005, submitted to WHO by CIEL).

Strachan WMJ, Huneault H, Schertzer WM, & Elder FC (1980) Organochlorines in precipitation in the Great Lakes region. In: Afghan BK & MacKay D ed., Hydrocarbons and halogenated hydrocarbons in the aquatic environment. New York, Plenum Press, pp. 387–396.

Sugiura K, Washino T, Hattori M, Sato E, & Goto M (1979) Accumulation of organochlorine compounds in fishes. Difference of accumulation factors by fishes. Chemosphere 6: 359–364.

Sunol C, Tusell JM, Gelp E, & Rodriguez-Farre E (1988) Convulsant effect of lindane and regional brain concentration of GABA and dopamine. Toxicology 49(2/3): 247–252.

Suter P, Horst K, Luetkemeier H, Herbst M, Terrier C, Lind H, & Ellgehausen H (1983) Three months toxicity study in rats with lindane: Part 1 & 2. Itingen, Research and Consulting Company AG (Unpublished report Celamerck No. 111AA-433-007/009, submitted to WHO by CIEL).

Suzuki M, Yamato Y, & Watanabe T (1975) Persistence of BHC (1,2,3,4,5,6-hexachlorocyclohexane) and dieldrin residues in field soils. Bull Environ Contam Towicol 14(5): 520–529.

Sweeney RA (1969) Metabolism of lindane by unicellular algae. In: Proceedings of the 12th Conference on Great Lakes Research, International Associationforf Great Lakes Research, pp. 98–102.

Szymczynski GA, Waliszewski SM, Tuszewski M, & Pyda P (1986) Chlorinated pesticides levels in human adipose tissue in the district of Poznan. J Environ Sci Health A21(1): 5–14.

Takabatake E (1978) Levels of organochlorine compounds and heavy metals in tissues of the general population in Japan. Geogr Med 8: 1–23.

Tanabe S, Tatsukawa R, Kawano M, & Hidaka H (1982) Global distribution and atmospheric transport of chlorinated hydrocarbons: HCH(BHC) isomers and DDT compounds in the Western Pacific, Eastern Indian, and Antarctic Oceans. J Oceanogr Soc Jpn 38(3): 137–148.

Tanabe S, Tanaka H, & Tatsukawa R (1984) Polychlorobiphenyls, Σ-DDT, and hexachlorocyclohexane isomers in the Western North Pacific ecosystem. Arch Environ Contam Toxicol 13: 731–738.

Tanaka K, Kurihara N, & Nakajima M (1977) Pathways of chlorophenol formation in oxidative biodegradation of BHC. Agric Biol Chem 41(4): 723–725.

Tanaka K, Kurihara N, & Nakajima M (1979) Oxidative metabolism of lindane and its isomers with microsomes from rat liver and housefly abdomen. Pestic Biochem Physiol 10: 96–103.

Thorpe E & Walker AIT (1973) The toxicology of dieldrin (HEOD). II. Comparative long-term oral toxicity studies in mice with dieldrin, DDT, phenobarbitone, β-BHC and γ-BHC. Food Cosmet Toxicol 11: 433–442.

Tolot F, Longlet JP, & Berthelon J (1969) Polyneurite due au lindane. J Méd Lyon 50: 747–754.

Tomczak S, Baumann K, & Lehnert G (1981) Occupational exposure to hexachlorocyclohexane. IV. Sex hormone alterations in HCH-exposed workers. Int Arch Occup Environ Health 48: 283–287.

Tooby TE & Durbin FJ (1975) Lindane residue accumulation and elimination in rainbow trout (*Salmo gairdnerii* Richardson) and roach (*Rutilus rutilus* L.). Environ Pollut 8(2): 79-89.

Trautmann A & Streit B (1979) [Sorption of lindane (γ-hexachlorocyclohexane) in Nitzschia actinastroïdes (Lemm.) v. Goor (dimethoate) at different growth conditions]. Arch Hydrobiol Suppl 55(3/4): 324–348 (in German).

Trifonova TK, Gladenko IN, & Schukla WD (1970) [Effect of γ--BHC and Sevin on reproduction]. Veterinarja 47: 91–93 (in Russian).

Trosko JE (1982) Inhibition of cell-to-cell communication by tumour promoters. Carcinogenesis 7: 565–585.

Truhaut R (1954) Communication to the International Symposium on Prevention of Cancer, São Paulo, 1954. In: Evaluation of the toxicity of pesticide residues in food. Report of a Joint Meeting of the FAO Committee on Pesticides in Agriculture and te WHO Expert Committee on Pesticide Residues, Geneva, 30 September–7 October 1963. Rome, Food and Agriculture Organization of the United Nations, p. 48 (FAO Meeting Report No. PL/1963/13. WHO/Food Add., 23 (1964)).

Tsukano Y (1973) Factors affecting disappearance of BHC isomers from rice field soil. Jpn Agric Res Q 7(2): 93–97.

Tsushimoto G, Chang CC, Trosko JE, & Matsumura F (1983) Cytotoxic, mutagenic, and cell-cell communication inhibitory properties of DDT, lindane, and chlordane on Chinese hamster cells *in vitro*. Arch Environ Contam Toxicol 12: 721–730.

Tu CM (1975) Interaction between lindane and microbes in soils. Arch Microbiol 105(2): 131–134.

Tu CM (1976) Utilization and degradation of lindane by soil microorganisms. Arch Microbiol 108: 259–263.

Tuinstra LGMT (1971) Organochlorine insecticide residues in human milk in the Leiden region. Neth Milk Dairy J 25: 24–32.

Turner JC (1979) Transplacental movement of organochlorine pesticide residues in desert bighorn sheep. Bull Environ Contam Toxicol 21: 116–124.

Turner JC & Shanks V (1980) Absorption of some organochlorine compounds by the rat small intestine—*in vivo*. Bull Environ Contam Toxicol 24: 652–655.

References

Turtle EE, Taylor A, Wright EN, Thearle RJP, Egan H, Evans WH, & Soutar NM (1963) The effects on birds of certain chlorinated insecticides used as seed dressings. J Sci Food Agric 14: 567–577.

Tussel JM, Sunol C, Gelpi E, & Rodriguez-Farre E (1988) Effect of lindane at repeated low doses. Toxicology 49: 375–379.

Tzoneva-Maneva MT, Kaloyaneva F, & Georgieva V (1971) Influence of diazinon and lindane on the mitotic activity and the caryotype of human lymphocytes, cultivated in vitro. Bibl Haematol 38(1): 344–347.

Uchida M, Kurihara N, Fujita T, & Nakajima M (1974) BHC isomers and related compounds. VII. Inhibitory effects of BHC isomers on sodium-potassium ion-dependent ATPase, yeast growth, and nerve conduction. Pestic Biochem Physiol 4(3): 260–265.

Ukeles R (1962) Growth of pure cultures of marine phytoplankton in the presence of toxicants. Appl Microbiol 10: 532–537.

Ullmann L, Sacher R, & Porrizello T (1986a) Primary skin irritation study with lindane in rabbits (4-hour occlusive application). (Unpublished report Celamerck No 111AD-465-003, submitted to WHO by CIEL).

Ullmann L, Claire R, & Bognar G (1986b) Test for delayed contact hypersensitivity in the albino guinea pig with lindane (maximization test). Itingen, Research and Consulting Company AG (Unpublished report No. 061650, Celamerck No. 111AC-467-001, submitted to WHO by CIEL).

Ullmann L, Mohler H, & Gembardt G (1986c) Four-hour acute aerosol inhalation toxicity study with lindane in rats. Itingen, Research and Consulting Company AG (Unpublished report No. 061637, submitted to WHO by CIEL).

Ullmann L, Claire R, & Bognar G (1987a) Contact hypersensitivity to Nexit Fluessig (CME 11129) in albino guinea pigs (maximization test). Itingen, Research and Consulting Company AG (Unpublished report Celamerck No. 11129-467-004, submitted to WHO by CIEL).

Ullmann J, Claire R, & Bognar G (1987b) Contact hypersensitivity to Nexit Stark (CME 11125) in albino guinea pigs (maximization test). Itingen, Research and Consulting Company AG (Unpublished report Celamerck No. 11125-467-002, submitted to WHO by CIEL).

Ullmann J, Claire R, & Bognar G (1987c) Contact hypersensitivity to Agronex Saatgutpuder (CME 11102) in albino guinea pigs (maximization test). Itingen, Research and Consulting Company AG (Unpublished report Celamerck No. 11102-467-003, submitted to WHO by CIEL).

Umweltbundesamt (1988–89) [Data on the environment 1988/89. Berlin, Schmidt E. Verlag, pp. 403, 410–412, 416–417, 521–522, 524, 526 (in German).

Uphouse L (1987) Decreased rodent sexual receptivity after lindane. Toxicol Lett 39: 7–14.

US Environmental Protection Agency (1976) National Interim Primary Drinking Water Regulations, Subpart B—Maximum contaminant levels. Washington, DC, Office of Water Supply (Report No. EPA-570/9-76-003).

US Environmental Protection Agency (1977) Notice of rebuttable presumption against registration and continued registration of pesticide products containing lindane. Fed Reg 42(33): 9816–9947.

US Environmental Protection Agency (1980) Lindane position document 2/3. Washington, DC, Office of Pesticide Programs.

US Environmental Protection Agency (1984) Health effects assessment for lindane. Cincinnatti, Ohio (Report No. EPA/540/1-86/056).

US Environmental Protection Agency (1985) Guidance for the reregistration of pesticide products containing lindane as the active ingredient. Washington, DC (Report No. RS-85).

US National Cancer Institute (1977) Bioassay of lindane for possible carcinogenicity. Washington, DC, US Department of Health, Education, and Welfare (Technical Report Series No. 14).

van Velsen FL (1986) The oestrogenicity of the environmental contaminant β-hexachlorocyclohexane. University of Utrecht (Thesis).

van Velsen FL, Franken MAM, van Leeuwen FXR, & Loeber JG (1984) [Study of semi-chronic oral toxicity of γ-HCH in the rat]. Bilthoven, National Institute of Public Health and Environmental Hygiene (Unpublished report No. 618209001) (in Dutch).

Vesselinovitch SD & Carlborg FW (1983) Lindane bioassay studies and human cancer risk assessment. Toxicol Pathol **11**(1): 12–22.

Vetter H, Kampe W, & Ranfft, K (1983) [Results of three years of comparative tests on vegetables, fruit and bread from modern and alternative producers]. Darmstadt, Association of German Institutes for Agricultural Testing and Research, pp. 1–11, 143 (VDLUFA Report No. 7) (in German).

Vodopick H (1975) Erythropoietic hypoplasia after exposure to γ-benzene hexachloride. J Am Med Assoc **234**(8): 850–851.

Voerman S & Besemer AFH (1970) Residues of dieldrin, lindane, DDT, and parathion in a light sandy soil after repeated application throughout a period of 15 years. J Agric Food Chem **18**(4): 717–719.

Voerman S & Besemer AFH (1975) Persistence of dieldrin, lindane, and DDT in a light sandy soil and their uptake by grass. Bull Environ Contam.Toxicol **13**(4): 501–505.

Vogel SM, Joy RM, & Narahashi T (1985) Lindane exerts both presynaptic and postsynaptic actions at frog neuromuscular junction. In: 15th Annual Meeting of the Society for Neuroscience, Part 2. Dallas, Texas, 20–25 October 1985. Soc Neurosci Abstr **11**: 1004.

Vohland HW, Portig J & Stein K (1981) Neuropharmacological effects of isomers of hexachlorocyclohexane. Toxicol Appl Pharmacol **57**(3): 425–438.

Vonk JW & Quirijns JK (1979) Anaerobic formation of α-hexachlorocyclohexane from γ-hexachlorocyclohexane in soil and by *Escherichia coli*. Pestic Biochem Physiol **12**: 68–74.

de Vos RH, van Dokkum W, Olthof PDA, Quirijns JK, Muys T, & van der Poll JM (1984) Pesticides and other chemical residues in Dutch total diet samples (June 1976–July 1978). Food Chem Toxicol **22**(1): 11–21.

Vukavic T, Pavkov S, Cusic S, Roncevic N, Vojinovic M, & Tokovic B (1986) Pesticide residues in human colostrum: seasonal variations, Yugoslavia. Arch Environ Contam Toxicol **15**: 525–528.

Wahid PA & Sethunathan N (1979) Sorption and desorption of α, β, and γ isomers of hexachlorocyclohexane in soils. J Agric Food Chem **27**(5): 1050–1053.

Wahid PA & Sethunathan N (1980) Sorption and desorption of lindane by anaerobic and aerobic soils. J Agric Food Chem **28**: 623–625.

References

Wammes JY, Wegman RCC, Hofstee AWM, Janssens H, Rieffe DW, Marsman JA, Groenemeijer GS, & van den Broek HH (1983) [Investigation into the presence of pesticides and related compounds in surface waters]. Bilthoven, National Institute of Public Health and Environmental Hygiene (Unpublished report No. 218102003) (in Dutch).

Wang HH & Grufferman S (1981) Aplastic anemia and occupational pesticide exposure: a case-control study. J Occup Med 23(5): 364–366.

Ware GW & Naber EC (1961) Lindane in eggs and chicken tissues. J Econ Entomol 54(4): 675–677.

Wassermann M, Ron M, Bercovici B, Wassermann D, Cucos S, & Pines A (1982) Premature delivery and organochlorine compounds. Polychlorinated biphenyls and some organochlorine insecticides. Environ Res 28: 106–112.

Wedberg JL, Moore S, III, Amore FJ, & McAvoy H (1978) Residues in food and feed. Organochlorine insecticide residues in bovine milk and manufactured milk products in Illinois, 1971–76. Pest Monit J 11(4): 161–164.

Wegman RCC & Greve PA (1980) Halogenated hydrocarbons in Dutch water samples over the years 1969–1977. In: Afghan BK & MacKay D (eds.), Hydrocarbons and halogenated hydrocarbons in the aquatic environment. New York, Plenum Press, pp. 405–415.

Weigert P, Müller J, Hans R, & Zufelde KP (1983) [Pesticide residues in foodstuffs. Communication I: Organochlorine compounds]. Berlin, Dietrich Reimer Verlag (ZEBS-Berichte 3) (in German).

Weisse I & Herbst M (1977) Carcinogenicity study of lindane in the mouse. Toxicology 7: 233–238.

Wellborn TL, Jr (1971) Toxicity of some compounds to striped bass fingerlings. Progr Fish-Cult 33: 32–36.

West I (1967) Lindane and haematologic reactions. Arch Environ Health 15: 97–101.

Wheatley GA (1965) The assessment and persistence of residues of organochlorine insecticides in soils and their uptake by crops. Proc Assoc Appl Biol 55: 325–329.

Whitehead CC, Downing AG, & Pettigrew RJ (1972a) The effects of lindane on laying hens. Br Poult Sci 13: 293–299.

Whitehead CC, Downie JN, & Phillips JA (1972b) BHC not found to reduce the shell quality of hen's eggs. Nature 239: 411–412.

Whitehead CC, Downie JN, & Phillips JA (1974) Some characteristics of the egg shells of quail fed gamma-BHC. Pestic Sci 5: 275–279.

WHO/FAO (1975) Data sheets on pesticides: No. 1- Lindane. Geneva (Unpublished document VBC/DS/75.12).

WHO (1985) Specifications for pesticides used in public health: insecticides-molluscicides-repellents—methods, 6th ed., Geneva.

WHO (1990) The WHO recommended classification of pesticides by hazard and guidelines to classification 1990–1991. Geneva (Unpublished document WHO/PCS/90.1 Rev. 1).

Wilkes LC, Tapprich BL, & Mulkey NS (1987a) Determination of ^{14}C-residues following oral administration of ^{14}C-lindane to lactating goats (in vivo and in vitro studies). Monument, California, Development Corporation (Unpublished report No. ADC 957a, submitted to WHO by CIEL).

Wilkes LC, Mulkey NS, Hallenbeck SA, Piznik M, & Wargo JP (1987b) Metabolism study of ^{14}C-lindane fed or topically applied to lactating goats. Monument, California, Development Corporation (Unpublished report No. ADC 957b, submitted to WHO by CIEL).

Williams DT, Benoit, FM, McNeil EE, & Otson R (1978) Organochlorine pesticide levels in Ottawa drinking water, 1976. Pestic Monit J **12**(3): 163.

Wirth H (1985) The adsorbtion of ^{14}C-labelled γ-HCH (lindane) at levels in the range of ng/litre to soil components. Geesthacht, GKSS Research Institute (Report No. GKSS 85/E/30).

Wittlinger R & Ballschmiter K (1987) Global baseline pollution studies. XI: Congener specific determination of polychlorinated biphenyls (PCBs) and occurrence of alpha- and gamma-hexachlorocyclohexane (HCH), 4.4'-DDE and 4.4'-DDT in continental air. Chemosphere **16**(10-12): 2497–2513.

Wolfe GW & Ralph JA (1980) Acute oral toxicity study in B6C3F1 mice: Lindane. Vienna, Virginia, Hazleton Laboratories America Inc. (Unpublished report submitted to WHO by CIEL).

Wolff G (1986) Maternal influence on phenotypic differentiation of a mutant mouse susceptible to neoplasia and obesity. In: Genetic toxicology of environmental chemicals, Part B, Genetic effects and applied mutagenesis. New York, Alan R. Liss Inc., pp. 33–38.

Wolff GL & Morrissey RL (1986) Increased responsiveness of lean pseudoagouti Avy/a female mice to lindane enhancement of lung and liver tumorigenesis. Proc Am Assoc Cancer Res **26**: 138.

Wolff GL, Roberts DW, & Galbraith DB (1986) Prenatal determination of obesity, tumour susceptibility and coat colour pattern in viable yellow (Avy/a) mice. J Hered **77**: 151–158.

Woodard F & Hagan EC (1947) Toxicological studies on the isomers and mixtures of isomers of benzenehexachloride. Fed Proc **6**: 386.

Yamaguchi I, Matsumura F, & Kadous AA (1979) Inhibition of synpatic ATPase by heptachlor epoxide in rat brain. Pestic Biochem Physiol **11**: 285–293.

Yamato Y, Kiyonaga M, & Watanabe T (1983) Comparative bioaccumulation and elimination of HCH-isomers in short-necked clam (*Venerupis japonica*) and guppy (*Poecilia reticulata*). Bull Environ Contam Toxicol **31**: 352–359.

Yoshida T & Castro TF (1970) Degradation of gamma-BHC in rice soils. Soil Sci Soc Am Proc **34**: 440–442.

Yule WN, Chiba M, & Morley HV (1967) Fate of insecticide residues. Decomposition of lindane in soil. J Agric Food Chem **15**(6): 1000–1004.

Zeilmaker MJ & Yamasaki H (1986) Inhibition of junctional intercellular communication as a possible short-term test to detect tumour promoting agents. Results with nine chemicals tested by dye-transfer assay in Chinese hamster V79 cells. Cancer Res **46**: 6180–6186.

Zesch A, Nitzsche K, & Lange M (1982) Demonstration of the percutaneous resorption of a lipophilic pesticide and its possible storage in the human body. Arch Dermatol Res **273**: 43–49.

RESUME ET EVALUATION; CONCLUSIONS; RECOMMANDATIONS

I. Résumé et évaluation

1.1 Propriétés générales

L'hexachlorocyclohexane technique (HCH) est composé de 65-70% d'alpha-HCH, de 7-10% de beta-HCH, de 14-15% de gamma-HCH, et d'environ 10% d'autres isomères et composés. Le lindane contient > 99% de gamma-HCH. C'est un solide, avec une tension de vapeur faible, peu soluble dans l'eau mais très soluble dans les solvants organiques comme l'acétone, et dans les solvants aromatiques et chlorés. Le coefficient de partage n-octanol/eau (log P_{oe}) est de 3,2-3,7.

Le lindane peut être dosé à part des autres isomères du HCH après extraction par partage liquide/liquide, chromatographie sur colonne et chromatographie en phase gazeuse avec détection par capture d'électrons. Extrêmement sensibles, ces méthodes analytiques permettent d'identifier des résidus de lindane de l'ordre du nanogramme par kilogramme ou par litre.

Depuis le début des années 50, le lindane est utilisé comme insecticide à large spectre; le secteur agricole y a recours notamment pour traiter les semences et les sols; on l'utilise pour traiter les arbres, le bois de construction et les matériaux entreposés; on l'applique sur le pelage des animaux pour éliminer les ectoparasites; on en fait également usage dans le domaine de la santé publique.

1.2 Transport, répartition et transformation dans l'environnement

Le lindane est fortement adsorbé aux sols contenant une grande quantité de matière organique; en outre, il peut s'imprégner dans le sol à la faveur d'une chute de pluie ou d'une irrigation artificielle. Sous l'effet des températures élevées des régions tropicales, il se dissipe en grande partie par volatilisation.

Le lindane subit une dégradation rapide (déchloration) sous l'action des rayons ultra-violets, pour former des pentachlorocyclohexènes (PCCHs) et des tetrachlorocyclohexènes (TCCHs). Lorsque le lindane se dégrade en sol humide ou inondé ou en plein champ, sa demi-vie peut aller de quelques jours à trois ans, selon le type de sol, le climat, la profondeur à laquelle il se

Résumé

trouve, etc. Dans les sols consacrés aux cultures que l'on trouve habituellement en Europe, sa demi-vie est de 40 à 70 jours. Les sols non-stériles permettent une biodégradation plus rapide du lindane que les sols stériles. Les conditions anaérobies sont les plus propices à sa métabolisation microbienne. Le lindane présent dans l'eau se dégrade sous l'action des microorganismes contenus dans les sédiments, pour former les mêmes produits.

Le lindane et les gamma-PCCHs sont fixés en quantité limitée par les plantes où ils subissent une translocation, surtout lorsque les sols contiennent une forte proportion de matières organiques. On trouve les résidus essentiellement dans les racines des plantes; la translocation ne s'effectue que peu—ou pas du tout—dans les tiges, les feuilles ou les fruits. La bioconcentration est rapide chez les microorganismes, les invertébrés, les poissons, les oiseaux et l'homme, mais la biotransformation et l'élimination le sont également lorsque l'exposition est interrompue. Les organismes aquatiques le fixent en plus grande quantité à partir de l'eau qu'à partir de la nourriture. Les facteurs de bioconcentration dans les organismes aquatiques vont de 10 à 6000 en laboratoire, et de 10 à 2600 sur le terrain.

1.3 Concentration dans l'environnement et exposition humaine

On a trouvé du lindane en suspension dans l'air au-dessus des océans à des niveaux de concentration de 0,039–0,68 ng/m^3; son niveau de concentration a atteint 1 ng/m^3 dans l'air de certains pays. Les concentrations dans les eaux de surface de nombreux pays européens étaient la plupart du temps inférieures à 0,1 µg/litre. Dans le cas du Rhin et de ses affluents, le niveau de concentration variait entre 0,01 et 0,4 µg/litre de 1969 à 1974; après 1974, il était inférieur à 0,1 µg/litre. Dans l'eau de mer, des concentrations de 0,001–0,002 µg/litre ont été enregistrées. Les concentrations de lindane dans le sol sont généralement faibles—de l'ordre de 0,001–0,01 mg/kg, sauf dans les zones de décharge.

On a découvert du gamma-HCH chez les poissons et les crustacés, à des niveaux de concentration allant de 'indécelable' à 2,5 mg/kg (calculés par rapport aux matières grasses) selon qu'ils vivent en eau douce ou dans l'eau de mer et en fonction de leur teneur—faible ou élevée —en matières grasses. Des valeurs d'environ 330 et 440 µg/kg (poids humide) ont été trouvées dans le tissu adipeux des ours polaires respectivement en 1982 et 1984. La concentration de lindane dans le foie des oiseaux prédateurs variait entre 0,01 et 0,1 mg/kg. En 1972–73, on a mesuré dans des oeufs d'épervier, en République fédérale d'Allemagne, des valeurs allant de 0,6 mg/kg à 11,1 mg/kg (calculées par rapport aux matières grasses).

Les niveaux de concentration de lindane dans l'eau potable sont généralement inférieurs à 0,001 µg/litre, et dans les pays industrialisés, c'est dans la nourriture que se trouve 90% du lindane absorbé par l'homme. Au cours des 25 dernières années, on a analysé dans bon nombre de pays certains produits alimentaires à la recherche de leur teneur en lindane: dans les céréales, les fruits, les légumes, les légumes à gousse, les huiles végétales, les concentrations allaient de 'indécelable' à 5 mg/kg de produit. En ce qui concerne le lait, les matières grasses animales, la viande et les oeufs, les concentrations allaient de 'indécelable' à 5,1 mg/kg de produit (calculés par rapport aux matières grasses). On a rarement trouvé des concentrations plus élevées. Les niveaux de concentration dans le poisson étaient généralement très inférieurs à 0,05 mg/kg de produit (calculés par rapport aux matières grasses). Des études de ration totale et de panier de la ménagère pour estimer l'apport quotidien de lindane chez l'homme, ont révélé d'importantes variations en fonction des époques: vers 1970, l'apport quotidien était égal ou inférieur à 0,05 µg/kg de poids corporel; en 1980, par contre, cet apport a diminué pour n'être plus qu'égal ou inférieur à 0,003 µg/kg. Aux USA, l'apport journalier de gamma-HCH entre 1976 et 1979 a diminué, passant de 0,005 à 0,001 µg/kg pour les tout-petits et de 0,01 à 0,006 µg/kg pour les enfants.

Dans un certain nombre de pays, on a déterminé la concentration en lindane dans des tissus d'individus appartenant à la population générale. Aux Pays-Bas, la teneur du sang était de l'ordre de < 0,1–0,2 µg/litre, mais des concentrations bien supérieures ont été trouvées dans le sang d'individus vivant dans des pays utilisant du HCH technique. Dans différents pays, on a mesuré des niveaux moyens de concentrations dans le tissu adipeux de l'homme allant de < 0,01 à 0,2 mg/kg (calculés par rapport aux matières grasses). Dans le lait humain, les concentrations de lindane sont généralement plutôt faibles, avec des moyennes allant de moins de 0,001 à 0,1 mg/kg, rapportées aux matières grasses; toutefois, on a constaté que ces valeurs s'abaissaient à la longue.

Ainsi, le lindane est présent partout dans le monde; on le retrouve dans l'air, l'eau, le sol, les sédiments, les organismes aquatiques et terrestres, ainsi que dans la nourriture, encore que les concentrations dans ces compartiments du milieu soient généralement faibles et diminuent peu à peu. L'homme est exposé quotidiennement, par les aliments qu'il consomme; c'est ainsi que l'on peut trouver du lindane dans le sang, les tissus adipeux, et dans le lait de femme; toutefois, les doses ingérées sont également en diminution.

Résumé

1.4 Cinétique et métabolisme

Chez les rats, le lindane est absorbé rapidement au niveau des voies digestives et se répartit dans l'ensemble des organes et tissus en quelques heures. C'est dans les tissus adipeux et la peau que l'on trouve les niveaux de concentration les plus élevés; diverses études ont fait état d'un rapport teneur des graisses/teneur du sang égal à 150–200, d'un rapport foie/sang de 5,3–9,6, et d'un rapport cerveau/sang de 4–6,5. Un rapport graisses/sang identique a été trouvé chez des rats ayant inhalé du lindane. Ces rapports, qui varient en fonction du sexe, sont supérieurs chez les femelles. Au niveau de la peau, la résorption s'effectue très lentement et dans une faible proportion, ce qui explique la faible toxicité du lindane après exposition cutanée.

Le lindane est métabolisé essentiellement dans le foie selon quatre réactions enzymatiques: la déshydrogénation en gamma-HCH, la déshydrochloration en gamma-PCCH, la déchloration en gamma-TCCH et l'hydroxylation en hexachlorocyclohexanol. Les produits finals après biotransformation sont des dérivés di-, tri-, tetra-, penta-, et hexachlorés. Ces métabolites sont excrétés essentiellement dans les urines sous forme libre ou conjugués à l'acide glucuronique, à l'acide sulfurique ou à la *N*-acétylcystéine. Le processus d'élimination est relativement rapide, avec des demi-vies de 3–4 jours chez le rat. Les bactéries et les champignons métabolisent le lindane en TCCH et en PCCH. La vitesse de la transformation métabolique dans les plantes est faible, la voie de dégradation principale passant par le PCCH pour aboutir au tetrachlorophenol, à des conjugués avec le bêta-glucose et à d'autres composés inconnus. Rien ne prouve que le lindane s'isomérise en alpha-HCH.

1.5 Effets sur les êtres vivants dans leur milieu naturel

Le lindane n'est pas très toxique pour les bactéries, les algues, les protozoaires: la dose sans effet est généralement de 1 mg/litre. Son action sur les champignons est variable, avec des doses sans effet variant entre 1 et 30 mg/litre, selon l'espèce. Il est modérément toxique pour les invertébrés et les poissons, les valeurs $CL(E)_{50}$ pour ces organismes étant de 20–90 µg/litre. Les études à court et à long terme portant sur trois espèces de poisson ont révélé que la dose sans effet est de 9 µg/litre; des concentrations de 2,1–23,4 µg/litre n'ont aucune conséquence sur la reproduction. Les valeurs de la CL_{50} pour les crustacés d'eau douce et marins varient entre 1 et 1100 µg/litre. Chez *Daphnia magna*, il y a réduction, fonction de la dose, du taux de reproduction; la dose sans effet se situe entre 11 et 19 µg/litre. Une dose de 1 mg/litre n'a eu aucun effet néfaste sur la reproduction des mollusques.

La DL_{50} pour les abeilles domestiques est de 0,56 µg/abeille.

Les DL_{50} orales aiguës pour un certain nombre d'espèces d'oiseaux se situent entre 100 et 1000 mg/kg de poids corporel. Des études à courte terme sur les oiseaux ont permis d'établir que des doses de 4–10 mg/kg de nourriture n'ont aucun effet, pas même sur la qualité de la coquille des oeufs. On a néanmoins constaté une moindre ponte chez les canes exposées à des doses de lindane allant jusqu'à 20 mg/kg de poids corporel.

Des chauves-souris sont mortes dans les 17 jours qui ont suivi leur exposition à des copeaux sur lesquels on avait appliqué du lindane à la dose indiquée; les copeaux en contenaient initialement 10–866 mg/m². On ne possède aucune donnée concernant les effets sur les individus et les écosystèmes.

1.6 Effets sur les animaux de laboratoire et effets in vitro

La toxicité orale aiguë du lindane est modérée: la DL_{50} pour les souris et les rats est de l'ordre de 60–250 mg/kg de poids corporel, selon le véhicule utilisé. La LD_{50} dermique pour les rats est d'environ 900 mg/kg de poids corporel. La toxicité est révélée par des signes d'excitation au niveau du système nerveux central.

Le lindane ne provoque aucune irritation ou sensibilisation de la peau; il est légèrement irritant pour les yeux.

Une étude de 90 jours sur des rats a permis d'établir à 10 mg/kg de nourriture (soit 0,5 mg/kg de poids corporel) la dose sans effet. A 50 et 250 mg/kg de nourriture, il y avait augmentation du poids du foie, des reins et de la thyroïde; à 250 mg/kg de nourriture, on constatait une augmentation de l'activité enzymatique du foie. Cette augmentation de l'activité enzymatique accélère la dégradation du lindane et autres dérivés. Une autre étude de 90 jours sur des rats a montré qu'une dose de 4 mg/kg de nourriture (soit 0,2 mg/kg de poids corporel) pouvait être définie comme dose sans effet nocif; on a observé que des concentrations égales ou supérieures à 20 mg/kg de nourriture pouvaient avoir un effet toxique au niveau des reins et du foie. Une étude toxicologique à court-terme sur les souris n'a pas permis de définir la dose sans effet.

On a constaté aucun effet toxique après administration de lindane à des chiens, pendant 63 semaines, 15 mg/kg de nourriture (soit 0,6 mg/kg de poids corporel). Une étude toxicologique menée pendant 2 ans sur des chiens et au cours de laquelle un grand nombre de paramètres ont été mesurés, a permis d'établir que l'administration de doses égales ou supérieures à 50 mg/kg de nourriture (soit 2 mg/kg de poids corporel) ne provoquait aucune anomalie apparente liée au traitement. Chez les animaux

Résumé

auxquels on avait administré des doses de 100 mg/kg de nourriture, on constatait toutefois une hausse des niveaux de phosphatase alcaline; avec des doses de 200 mg/kg, on observait des anomalies du tracé électro-encéphalographique, indiquant une excitation neuronale aspécifique.

Chez des rats ayant inhalé du lindane à raison de 0,02–4,54 mg/m^3, 6 heures par jour pendant 3 mois, la dose la plus forte a entraîné une élévation des valeurs du cytochrome P450 hépatique; la dose sans effet nocif a été fixée à 0,6 mg/m^3. Lors de deux études au long cours menées sur des rats de nombreuses années auparavant, on a expérimenté des doses de 10–1600 mg/kg de nourriture. L'une de ces études fixe à 50 mg/kg de nourriture la dose sans effet nocif. A 100 mg/kg de nourriture, le foie augmente, une hypertrophie hépatocellulaire apparaît, ainsi qu'une dégénérescence graisseuse et une nécrose. L'autre étude établit à 25 mg/kg de nourriture (soit 1,25 mg/kg de poids corporel) la dose sans effet; en doublant cette dose, on note une hypertrophie hépatocellulaire et une dégénerescence graisseuse.

Des recherches ont été entreprises pour vérifier les effets que le lindane, après avoir été administré à des souris, des rats, des chiens et des porcs par voie orale, sous-cutanée et intrapéritonéale, pouvait avoir sur tous les aspects de la reproduction (chez les rats, sur trois générations), et évaluer sa toxicité pour l'embryon et sa tératogénicité. Le lindane administré par voie orale et parentérale n'a eu aucun effet tératogène (les côtes surnuméraires étant considérées comme des variations). Des doses de 10 mg/kg de poids corporel et plus administrées oralement—par gavage—se sont avérées toxiques pour le foetus et/ou la mère; on considère qu'une dose de 5 mg/kg de poids corporel est sans effet nocif. L'étude effectuée sur trois générations de rats, a montré que des doses de lindane allant jusqu'à 100 mg/kg de nourriture n'avaient d'influence ni sur la reproduction, ni sur la maturation; avec 50 mg/kg de nourriture, des modifications morphologiques intervenaient au niveau du foie parmi la progéniture de la troisième génération, preuve d'une induction enzymatique. La dose sans effet était dans ce cas de 25 mg/kg (soit 1,25 mg/kg de poids corporel).

Dans une étude de 22 jours effectuée sur des rats, on a déterminé que la dose sans effet neurotoxique était de 2,5 mg/kg de poids corporel.

Des études bien conçues ont été effectuées pour déterminer la mutagénicité du lindane. Selon les résultats des recherches très larges entreprises, le lindane ne peut en aucun cas provoquer des mutations génétiques chez les bactéries ni dans des cellules mammaliennes; il n'entraîne pas non plus, chez *Drosophila melanogaster*, des mutations récessives liées au sexe. D'autres expériences—*in vitro* et *in vivo*—ont montré qu'en outre le lindane ne provoque ni anomalies chromosomiques, ni échange de chromatides soeurs dans des cellules mammaliennes. La recherche de lésions de l'ADN bactérien et de liaisons covalentes avec l'ADN dans le foie de rats et de souris *in vivo*, après administration par voie orale, a également donné des résultats négatifs. Les rares études où des résultats positifs ont été obtenus péchaient soit par une mauvaise conception d'ensemble, soit par le fait que la pureté du composé étudié n'avait pas été précisée. Quoi qu'il en soit, on peut dire que le lindane n'a globalement aucun pouvoir mutagène.

Des études de cancérogénécité ont été effectuées chez la souris et le rat, avec des doses allant respectivement jusqu'à 600 mg/kg de nourriture et 1600 mg/kg de nourriture. Aux doses égales ou supérieures à 160 mg/kg de nourriture, on a observé des nodules hyperplasiques et/ou des adénomes hépatocellulaires chez les souris; lors de certaines études, les doses administrées ont dépassé le maximum toléré. Deux expériences ont montré qu'aucune élévation de l'incidence des tumeurs n'intervient lorsque l'on donne à des souris et à des rats des doses allant respectivement jusqu'à 160 et 640 mg/kg de nourriture.

Les résultats d'études sur l'initiation–promotion de la cancérogénicité, sur le mode d'action du lindane et sur sa mutagénicité indiquent que la réponse tumorigène observée avec le gamma-HCH chez la souris est sous la dépendance d'un mécanisme non génétique.

1.7 Effets sur l'homme

Le lindane a été à l'origine de plusieurs cas d'intoxications mortelles et non mortelles; il s'agissait soit d'accidents, soit d'absorption délibérée (suicide), soit d'une simple négligence (absence de précautions) ou d'utilisation impropre de produits médicaux contenant du lindane. Les symptômes consistaient en nausées, agitation, maux de tête, vomissements, tremblements, ataxie, convulsions toniques-cloniques et/ou modifications du tracé électroencéphalographique. Ces effets sont réversibles après interruption de l'exposition ou traitement symptomatique.

Depuis 40 ans, l'utilisation du lindane est très répandue; pourtant on cite peu de cas d'intoxications survenant dans le contexte professionnel. Chez les individus qui travaillent à la fabrication du lindane ou à son épandage, donc soumis à une exposition prolongée, on a seulement constaté une augmentation de l'activité des enzymes métabolisantes au niveau du foie. Rien ne prouve l'existence d'une quelconque relation—évoquée dans

Résumé

certaines publications—entre exposition au lindane et apparition d'anomalies hématologiques. Quelques études de toxicologie aiguë ou à courte terme chez l'homme indiquent qu'une dose d'environ 1,0 mg/kg de poids corporel ne provoque pas d'intoxication; en revanche, à la dose de 15–17 mg/kg de poids corporel, apparaissent de graves symptômes d'intoxication.

Appliqué sur la peau, le lindane est absorbé à hauteur d'environ 10%; la proportion est plus élevée s'il y a lésions.

2. Conclusions

2.1 Population générale

Le lindane circule dans l'environnement et il est présent dans les chaînes alimentaires; de ce fait, l'homme ne peut échapper à l'exposition. Toutefois, l'apport alimentaire quotidien et l'exposition totale de la population dans son ensemble diminuent peu à peu, et, nettement inférieurs à la dose journalière admissible (DJA) conseillée, ne constituent pas une menace sérieuse pour la santé publique.

2.2 Groupes de population particulièrement exposés

La présence de lindane dans le lait maternel expose les enfants nourris au sein, à des doses généralement inférieures à la DJA, donc non toxiques. Les niveaux d'exposition existants—que l'on les souhaiterait tout de même inférieurs—n'interdisent pas l'allaitement au sein.

En ce qui concerne l'utilisation thérapeutique du lindane pour traiter la gale et les poux de corps, il convient de se conformer strictement aux doses prescrites.

2.3 Exposition professionnelle

Manipuler du lindane ne présente aucun danger, à condition de prendre toutes les précautions indiquées pour éviter le plus possible l'exposition.

2.4 Effets sur l'environnement

Les chauves-souris, qui s'accrochent au bois traité avec du lindane aux doses indiquées, en subissent les effets toxiques. Exception faite des résultats d'étude concernant les déversements accidentels dans le milieu aquatique, rien ne permet d'affirmer que la présence de lindane dans l'environnement constitue un danger sérieux pour d'autres êtres vivants.

3. Recommandations

1. Afin de réduire au minimum la pollution de l'environnement par d'autres isomères de HCH, il convient d'utiliser le lindane (> 99% de gamma-HCH) au lieu du HCH technique.

2. Afin d'éviter la pollution de l'environnement, il faut adopter des solutions adéquates pour se débarrasser des sous-produits et des effluents provenant des usines de fabrication de lindane.

3. Il faut prendre garde à ce que les déchets de lindane ne polluent ni les sols, ni les eaux.

4. Il faut donner à ceux qui manipulent du lindane les indications nécessaires sur les méthodes d'application et les précautions d'utilisation.

5. Il faut effectuer des études de cancérogénécité au long cours, qui soient conformes aux normes actuelles.

6. Il faut poursuivre la surveillance de la dose de lindane quotidiennement absorbée par la population générale.

RESUMEN Y EVALUACIONES; CONCLUSIONES; RECOMENDACIONES

1. Resumen y evaluación

1.1 Propiedades generales

El hexaclorociclohexano (HCH) de calidad técnica está formado por un 65–70% de alfa-HCH, un 7–10% de beta-HCH, un 14–15% de gamma-HCH y aproximadamente un 10% de otros isómeros y compuestos. El lindano contiene más del 99% de gamma-HCH. Es un compuesto sólido, con baja presión de vapor y poco soluble en agua, pero muy soluble en disolventes orgánicos, como la acetona, y en disolventes aromáticos y clorados. El coeficiente de reparto n-octanol/agua (log P_{oa}) es de 3,2–3,7.

El lindano puede determinarse por separado de los demás isómeros del HCH tras su extracción por reparto líquido/líquido, cromatografía en columna y detección por cromatografía de gases con captura de electrones. Como estos métodos analíticos son sumamente sensibles, es posible identificar residuos de lindano del orden de nanogramos por kilogramo o por litro.

El lindano lleva utilizándose desde el comienzo los años 50 como insecticida de amplio espectro con fines agrícolas y de otro tipo, de los que cabe mencionar el tratamiento de semillas y de suelos, las aplicaciones en árboles, madera y materiales almacenados, el tratamiento de animales contra los ectoparásitos y en la salud pública.

1.2 Transporte, distribución y transformación en el medio ambiente

En los suelos con un alto contenido de materia orgánica se observa una intensa adsorción del lindano; además, puede penetrar en el suelo con el agua de la lluvia o del riego artificial. La volatilización parece ser una importante vía de dispersión en las elevadas temperaturas de las regiones tropicales.

El lindano experimenta una rápida degradación (descloración) por acción de los rayos ultravioleta, formando pentaclorociclohexenos (PCCH) y tetraclorociclohexenos (TCCH). Cuando el lindano se descompone en el medio ambiente en condiciones de humedad o inmersión y en condiciones de campo, su semivida varía de unos días a tres años, en función del tipo de suelo, del clima, de la profundidad a la que se haya aplicado y de otros factores. En los suelos agrícolas normales en Europa su semivida es de 40 a 70 días. La biodegradación del lindano es mucho más rápida en suelos no esterilizados que en los esterilizados. Las condiciones anaerobias son las más favorables para su metabolización microbiana. El lindano presente en el agua es degradado principalmente por microorganismos de los sedimentos para formar los mismos productos de degradación.

Las plantas absorben y translocan en su interior cantidades limitadas de lindano y de gamma-PCCH, especialmente en suelos con un elevado contenido de materia orgánica. Los residuos se depositan sobre todo en las raíces de las plantas, y son pocos o ninguno los que se desplazan a las ramas, las hojas o los frutos. En los microorganismos, los invertebrados, los peces, las aves y el hombre tiene lugar una bioconcentración rápida, pero cuando se interrumpe la exposición se biotransforman y eliminan en un tiempo relativamente breve. En los organismos acuáticos es más importante su absorción a partir del agua que de los alimentos. Los factores de bioconcentración de estos organismos en condiciones de laboratorio variaron desde un valor aproximado de 10 hasta 6000; en condiciones de campo oscilaron entre 10 y 2600.

1.3 Niveles medioambientales y exposición humana

En el aire oceánico se han encontrado concentraciones de lindano de 0,039–0,68 ng/m^3, y en el aire de algunos países se han medido cantidades de hasta 11 ng/m^3. Las concentraciones estimadas en aguas de superficie de varios países europeos fueron en general inferiores a 0,1 µg/litro. Su concentración en el río Rin y sus afluentes en el período 1969-74 osciló entre 0,01 y 0,4 µg/litro; después de 1974 se mantuvo por debajo de 0,1 µg/litro. En el agua marina se han detectado niveles de 0,001–0,02 µg/litro. Las concentraciones de lindano en el suelo son por lo general bajas, del orden de 0,001–0,01 mg/kg, excepto en zonas de vertido de basuras.

En pescados y mariscos se han detectado concentraciones de gamma-HCH que oscilan entre valores no detectables y 2,5 mg/kg (valores referidos a las grasas), dependiendo de que vivan en agua dulce o agua marina y de que su contenido en grasa sea alto o bajo. En el tejido adiposo de los osos polares se encontraron en 1982 y 1984 niveles aproximados de 330 y 440 µg/kg (peso húmedo) respectivamente. La concentración de

Resumen

lindano en el hígado de aves predadoras oscilaba entre 0,01 y 0,1 mg/kg. Los huevos de gavilán recogidos en 1972-73 en la República Federal de Alemania contenían entre 0,6 y 11,1 mg/kg (cálculo referido a las grasas).

Las concentraciones de lindano en el agua potable generalmente son inferiores a 0,001 µg/litro; en los países industrializados más del 90% de la ingestión humana de lindano procede de los alimentos. En los últimos 25 años se ha analizado el contenido de lindano de determinados productos alimenticios de un gran número de países. Las concentraciones halladas en cereales, frutas, hortalizas, legumbres y aceites vegetales variaron entre valores no detectables y 5 mg/kg de producto, y en la leche, las grasas, la carne y los huevos, entre valores no detectables y 5,1 mg/kg (referido a las grasas). Sólo en unos pocos casos se detectaron concentraciones más altas. Sus niveles en el pescado eran, en general, muy inferiores a 0,05 mg/kg de producto (referido a las grasas).

En estudios sobre dieta total y cesta de la compra para estimar la ingestión humana diaria de lindano, se observó una clara diferencia con el paso del tiempo: la ingestión en el período de alrededor de 1970 llegaba a 0,05 µg/kg de peso corporal al día, mientras que en 1980 esta cifra había descendido a 0,003 µg/kg de peso corporal al día o menos. En los Estados Unidos, la ingestión de gamma-HCH entre 1976 y 1979 disminuyó de 0,005 a 0,001 µg/kg de peso corporal al día en los lactantes y de 0,01 a 0,006 µg/kg de peso corporal al día en los niños de corta edad.

En algunos países se ha determinado el contenido de lindano en los tejidos corporales de la población general. En los Países Bajos, el contenido en la sangre era del orden de < 0,1-0, µg/litro, pero se hallaron concentraciones mucho más altas en varios países en los que se utilizaba HCH de calidad técnica. Las concentraciones medias en el tejido adiposo humano en distintos países varió entre < 0,01 y 0,2 mg/kg (referido a las grasas). La concentración de lindano en la leche humana suele ser bastante baja, con unos niveles medios que van desde < 0,001 hasta 0,1 mg/kilo (referido a las grasas); sin embargo, se ha producido una disminución manifiesta con el tiempo.

Así pues, el lindano se halla distribuido por todo el mundo y se puede detectar en el aire, el agua, el suelo, los sedimentos, los organismos acuáticos y terrestres y los alimentos, aunque las concentraciones en estos distintos compartimentos ambientales son en general bajas y están decreciendo progresivamente. El hombre está expuesto a diario por conducto de los alimentos, habiéndose detectado lindano en los tejidos sanguíneo y adiposo y en la leche materna; sin embargo, los niveles de ingestión también están disminuyendo.

1.4 Cinética y metabolismo

En las ratas, el lindano se absorbe rápidamente del tracto gastrointestinal y en unas horas se distribuye por todos los órganos y tejidos. Las concentraciones más elevadas se dan en el tejido adiposo y en la piel; en varios estudios, el cociente grasa:sangre era de alrededor de 150–200, el cociente hígado:sangre, 5,3–9,6 y el cociente cerebro:sangre, 4–6,5. El mismo cociente grasa:sangre se encontró en ratas expuestas por inhalación. Estos cocientes varían en función del sexo, siendo más elevados en las hembras. La absorción por la piel tras la aplicación cutánea de lindano es lenta y muy limitada; esto puede explicar la baja toxicidad del lindano después de la exposición cutánea.

El lindano se metaboliza sobre todo en el hígado mediante cuatro reacciones enzimáticas: deshidrogenación a gamma-HCH, deshidrocloración a gamma-PCCH, descloración a gamma-TCCH e hidroxilación a hexaclorociclohexanol. Los productos finales de la biotransformación son compuestos di-, tri-, tetra-, penta- y hexaclorados. Estos metabolitos se excretan fundamentalmente por la orina, en forma libre o conjugada con ácido glucurónico, ácido sulfúrico o N-acetilcisteína. La eliminación es relativamente rápida, con una semivida en ratas de 3 a 4 días. Las bacterias y los hongos metabolizan el lindano a TCCH y PCCH. La velocidad de transformación metabólica en las plantas es baja, y la vía de degradación más importante es a través del PCCH a tri- y tetraclorofenol y productos conjugados con beta-glucosa y otros compuestos desconocidos. No existen pruebas de la isomerización del lindano a alfa HCH.

1.5 Efectos en los seres vivos del medio ambiente

El lindano no es muy tóxico para las bacterias, las algas ni los protozoos: el nivel carente de efecto fue en general de 1 mg/litro. Su acción sobre los hongos es variable; los niveles sin observación de efectos fueron de 1 a 30 mg/litro, según las especies. Es moderadamente tóxico para los invertebrados y los peces, siendo los valores de la $C(E)L_{50}$ para esos organismos de 20–9 µg/litro. En estudios de corta y larga duración con tres especies de peces, el nivel sin observación de efectos fue de 9 µg/litro; no se observaron efectos en la reproducción con niveles de 2,1–23,4 µg/litro. Los valores de la CL_{50} para crustáceos dulceacuícolas y marinos variaron entre 1 y 1100 µg/litro. La inhibición de la reproducción de *Daphnia magna* dependía de la dosis; el nivel sin observación de efectos fue del orden de 11–19 µg/litro. No se observaron efectos adversos en la reproducción de moluscos con dosis de 1 mg/litro.

La DL_{50} para la abeja de la miel fue de 0,56 µg/abeja.

Resumen

Los valores de la DL$_{50}$ aguda por vía oral para varias especies de aves fueron de 100 a 1000 mg/kg de peso corporal. En estudios de corta duración con aves, las dosis de 4–10 mg/kg en la dieta no tuvieron efecto, ni siquiera sobre la calidad de la cáscara de los huevos. Sin embargo, en patas ponedoras tratadas con dosis de lindano de hasta 20 mg/kg de peso corporal disminuyó la producción de huevos.

Todos los murciélagos expuestos a virutas de madera con un contenido inicial de lindano de 10–866 mg/m^2, resultado de la aplicación de la dosis recomendada, murieron en un plazo de 17 días. No se obtuvieron datos acerca de los efectos en poblaciones y ecosistemas.

1.6 Efectos en los animales de experimentación e in vitro

La toxicidad aguda por vía oral del lindano es moderada: la DL$_{50}$ para ratones y ratas oscila entre 60 y 250 mg/kg de peso corporal, en función del vehículo utilizado. La DL$_{50}$ por vía cutánea en ratas es de aproximadamente 900 mg/kg de peso corporal. La toxicidad se manifestó en forma de estimulación del sistema nervioso central.

El lindano no irrita ni sensibiliza la piel; es ligeramente irritante para los ojos.

En un estudio de 90 días en ratas, la concentración máxima sin efecto fue de 10 mg/kg alimento (equivalente a 0,5 mg/kg de peso corporal). Con niveles de 50 y 250 mg/kg de alimento aumentaron los pesos del hígado, los riñones y el tiroides; con 250 mg/kg de alimento, se observó un aumento en la actividad enzimática del hígado. Este aumento acelera la degradación del lindano y de otros compuestos. En otro estudio de 90 días en ratas, se consideró que el nivel máximo sin efectos adversos era de 4 mg/kg de alimento (equivalente a 0,2 mg/kg de peso corporal); se observó toxicidad renal y hepática a concentraciones de 20 mg/kg y superiores. Un estudio de toxicidad de corta duración en ratones se consideró insuficiente para establecer la concentración sin efectos.

La administración de lindano a perros en dosis de 15 mg/kg de alimento (equivalentes a 0,6 mg/kg de peso corporal) durante 63 semanas no tuvo efectos tóxicos. En un estudio de dos años de duración sobre la toxicidad de este compuesto en perros, en el que se midió un gran número de parámetros, no se observaron anomalías relacionadas con el tratamiento con dosis de 50 mg/kg de alimento (equivalentes a 2 mg/kg de peso corporal) e inferiores. Sin embargo, en el grupo que recibió 100 mg/kg de alimento aumentó el nivel de fosfatasa alcalina; y con 200 mg/kg de alimento aparecieron anomalías electroencefalográficas indicativas de irritación neuronal inespecífica.

En ratas expuestas por vía respiratoria a concentraciones de lindano de 0,02–4,54 mg/m^3, 6 horas al día durante 3 meses, la dosis más alta indujo un incremento de los valores del citocromo P450 hepático; el nivel sin observación de efectos adversos fue de 0,6 mg/m^3. En dos estudios de larga duración en ratas, realizados hace muchos años, se ensayaron dosis de 10–1600 mg/kg de alimentos. En uno de estos estudios se determinó un nivel sin observación de efectos adversos de 50 mg/kg de alimento (equivalente a 2,5 mg/kg de peso corporal). Con 100 mg/kg de alimento se producía un aumento del peso del hígado, hipertrofia hepatocelular, degeneración grasas y necrosis. En el otro estudio, la dosis de 25 mg/kg de alimento (equivalente a 1,25 mg/kg de peso corporal) no tenía efectos, pero con 50 mg/kg de alimento se observaron signos de hipertrofia hepatocelular y degeneración grasas.

Se han investigado los efectos del lindano en todos los aspectos de la reproducción (en tres generaciones de ratas), y su embriotoxicidad y teratogenia tras la administración oral, subcutánea e intraperitoneal en ratones, ratas, perros y cerdos. No se observaron efectos teratogénicos tras la administración oral o parenteral (las costillas supernumerarias se consideraron variaciones). Se pusieron de manifiesto fetotoxicidad y/o efectos tóxicos maternos con dosis de 10 mg/kg de peso corporal y superiores administradas mediante sonda oral; se considera que el nivel sin efectos adversos es de 5 mg/kg de peso corporal. En el estudio de tres generaciones de ratas con dosis de hasta 100 mg/kg de alimentos el lindano no ejerció efecto alguno en la reproducción ni la maduración, pero con 50 mg/kg de alimento se produjeron cambios morfológicos en el hígado, que demostraban la inducción enzimática registrada en la descendencia de la tercera generación. El nivel sin observación de efectos en este ensayo fue de 25 mg/kg de alimento (equivalente a 1,25 mg/kg peso corporal).

En un estudio de 22 días en ratas se observó que la dosis sin efecto neurotóxico era de 2,5 mg/kg de peso corporal.

Se han hecho estudios suficientes sobre la mutagenicidad del lindano. En las amplias investigaciones realizadas sobre su capacidad para inducir mutaciones génicas en bacterias y células de mamíferos y para provocar mutaciones letales recesivas ligadas al sexo en *Drosophila melanogaster*, se obtuvieron siempre resultados negativos. El lindano

Resumen

también dio resultados negativos en los ensayos *in vitro* e *in vivo* realizados con células de mamíferos sobre lesiones cromosómicas e intercambio de cromátidas hermanas. Tambien fueron negativos los resultados de los ensayos para determinar las lesiones en el ADN de bacterias y los de las pruebas *in vivo* para observar la formación de enlaces covalentes con el ADN de hepatocitos de ratones y ratas tras su administración oral. En los escasos ensayos en los que se obtuvieron resultados positivos, el sistema de estudio no era adecuado o no se informó sobre la pureza del compuesto ensayado. Sin embargo, en conjunto, el lindano no parece tener potencial mutagénico.

Se han llevado a cabo estudios en ratones y ratas para determinar el potencial carcinogénico del lindano con dosis de hasta 600 mg/kg de alimento en ratones y de hasta 1600 mg/kg de alimento en ratas. En ratones que recibieron dosis de 160 mg/kg de alimento o superiores se observaron nódulos hiperplásicos y/o adenomas hepatocelulares; en algunos estudios, las dosis utilizadas superaban la máxima tolerada. En dos estudios en ratones con dosis de hasta 160 mg/kg de alimentos como máximo y uno en ratas con 640 mg/kg de alimentos no se vio ningún aumento en la incidencia de tumores.

Los resultados de los estudios sobre la iniciación y el estímulo de la carcinogenicidad, sobre el mecanismo de acción y sobre la mutagenicidad ponen de manifiesto que en la respuesta tumorigénica observada con el gamma-HCH en ratones interviene un mecanismo no genético.

1.7 Efectos en el ser humano

Se ha informado de varios casos de envenenamiento mortal y de enfermedad no mortal por lindano, producidos de manera accidental, intencionada (suicidio) o por una grave negligencia en las precauciones de seguridad o la utilización inadecuada de productos médicos con lindano. Los síntomas son náuseas, agitación, dolor de cabeza, vómitos, temblor, ataxia, convulsiones tónico-clónicas y/o cambios en el trazado electro-encefalográfico. Estos efectos eran reversibles tras la interrupción de la exposición o el tratamiento sintomático.

A pesar de su uso generalizado durante 40 años, se ha informado de muy pocos casos de envenenamiento en el trabajo. En los trabajadores expuestos durante largos períodos, en la fabricación o la aplicación del lindano, el único síntoma observado fue una mayor actividad de las enzimas hepáticas metabolizadoras de fármacos. No hay pruebas de la relación, sugerida en algunas publicaciones, entre la exposición al lindano y la aparición de anomalías hematológicas. Algunos estudios de toxicidad aguda y de corta duración en la especie humana indican que una dosis aproximada de 1,0 mg/kg de peso corporal no produce envenenamiento; sin embargo, con una dosis de 15–17 mg/kg de peso corporal se observaron síntomas de intoxicación grave.

Se absorbe alrededor del 10% de la dosis de aplicación cutánea, aunque a través de la piel lesionada pasa mayor cantidad.

2. Conclusiones

2.1 Población general

El lindano circula en el medio ambiente y está presente en las cadenas troficas, de manera que la especie humana seguirá estando expuesta. Sin embargo, la ingestión diaria y la exposición total de la población general están disminuyendo gradualmente; se encuentran claramente por debajo de la ingestión diaria admisible y no constituyen un problema para la salud pública.

2.2 Subpoblaciones especialmente expuestas

La presencia de lindano en la leche materna determina la exposición de los lactantes a niveles que generalmente son inferiores a la ingesta diaria admisible y que, por consiguiente, no son un problema para la salud. Aunque sería preferible que los niveles de exposición fueran inferiores, los actuales no representan un factor limitante de la práctica de la lactancia natural.

Se deben seguir rigurosamente las prescripciones en relación con el uso terapéutico del lindano contra la sarna y los piojos.

2.3 Exposición profesional

El lindano se puede manejar sin riesgo siempre que se observen las precauciones recomendadas para reducir al mínimo la exposición.

Resumen

2.4 Efectos en el medio ambiente

El lindano es tóxico para los murciélagos que reposan en estrecho contacto con madera tratada de acuerdo con las recomendaciones para la aplicación. Si se exceptúan los resultados obtenidos en los estudios sobre derrames en el medio acuático, no hay pruebas que indiquen que la presencia de lindano en el medio ambiente plantee un riesgo importante para las poblaciones de otros organismos.

3. Recomendaciones

1. A fin de reducir al mínimo la contaminación del medio ambiente por otros isómeros del HCH, se debe utilizar lindano (> 99% de gamma-HCH) en lugar de HCH de calidad técnica.

2. Con objeto de evitar la contaminación del medio ambiente, los subproductos y efluentes de la fabricación del lindano se deben eliminar de manera adecuada.

3. En la eliminación de lindano, hay que tomar precauciones para evitar la contaminación de las aguas naturales y del suelo.

4. Como en el caso de otros plaguicidas, las personas encargadas del manejo del lindano deben recibir instrucciones adecuadas acerca de la manera de aplicarlo.

5. Se deben realizar ensayos de carcinogenicidad de larga duración diseñados con arreglo a las normas actuales.

6. Se debe seguir vigilando la ingestión diaria de lindano por parte de la población general.